Seeking Societal Symbiosis for the Sustainable Use of Materials

'Crucially, this book bridges a gap in our understanding of how different societal sectors can be united to define roles, identify shared interests and devise strategies that complement progress towards more sustainable material use. This message will be of particular interest to the chemical industry as it seeks new pathways towards a cyclical economy.'

Dr Simon Vandestadt, *Director, Senior Executive and Researcher, Specialist in peace, conflict and sustainability, Pacific and Papua New Guinea*

We live in a material world – one with a rising population dependent upon a base of natural resources that is already seriously depleted and declining rapidly. This book considers the societal use of materials: historically, in the less-than-optimal present and aspirationally as a central thread in addressing wider sustainable development challenges. It accounts for linked chemical, physical and socio-economic consequences right through value chains, from raw material extraction through manufacture and use and onwards to post-use.

The author, Mark Everard, proposes that the four principal societal sectors – business, government, civil society organisations and knowledge-providers – though often currently dissipating effort by working antagonistically, can optimally work symbiotically around co-created and consensual long-term sustainability goals to accelerate necessary progress. Practical examples of 'baby steps' towards constructive collaboration within and between societal sectors are recognised, and lessons are drawn for how they can shape more committed and intentional collaboration for sustainable development.

Seeking Societal Symbiosis touches upon many aspects of science – from chemistry, chemical engineering and material science, through to construction/built environment, sustainability, regulatory science, resources economics and social science. It will be valuable reading for industry managers and executives, the European Commission and MEPs, other national regulators and legislators, university students and lecturers, and NGOs.

Seeking Societal Symbiosis for the Sustainable Use of Materials

Mark Everard

CRC Press
Taylor & Francis Group
Boca Raton London New York

CRC Press is an imprint of the
Taylor & Francis Group, an **informa** business

Designed Cover Image: Author's original

First edition published 2026
by CRC Press
2385 NW Executive Center Drive, Suite 320, Boca Raton FL 33431

and by CRC Press
4 Park Square, Milton Park, Abingdon, Oxon, OX14 4RN

CRC Press is an imprint of Taylor & Francis Group, LLC

© 2026 Mark Everard

Library of Congress Cataloging-in-Publication Data
[Insert LoC Data here when available]

ISBN: 9781041069232 (hbk)
ISBN: 9781041069225 (pbk)
ISBN: 9781003637875 (ebk)

DOI: 10.1201/9781003637875

Typeset in Times
by codeMantra

Contents

About the Author

Professor Mark Everard is a Visiting Professor at Bournemouth University and also an Associate Professor of Ecosystem Services at the University of the West of England (UWE Bristol). He also works as a consultant, broadcaster and author.

Mark is Vice President of the Institution of Environmental Sciences (IES), a Fellow of the Linnean Society, an Ambassador to the Angling Trust and to WaterHarvest, and a science advisor to WildFish (formerly Salmon & Trout Conservation UK), Tiger Water (India) and a range of other intergovernmental and national bodies. His work on sustainable development, ecosystem services, nature conservation and natural resource management has extended over a half-century and across five continents in both the developed and developing worlds.

Mark has worked directly on sustainability issues with the chemical sector for over a quarter century, developing challenges and strategies in partnership with businesses and trade associations that have traction with both practice and policy at an international scale. This advisory work in the chemical sector with governments and businesses has taken him from America to Australia, Africa and Asia as well as Europe.

In addition to his work in academia, formerly in government, with businesses and in the NGO sector, Mark is also a communicator with a wide portfolio of systems and sustainable development activities across a range of media including written publications (45 books and over 140 peer-reviewed scientific papers to date), social media and with broadcasters including frequent contributions to television and radio.

Introduction

1

Change is the new normal. Arguably though, change has been a constant throughout human history as we have responded to shifts in the environment, demography, knowledge, discoveries and innovations, belief systems, political ideologies, power relations within society and many more influences. Ingrained traditions, resource use patterns including tastes and taboos, market pressures, and legal and other restrictions as well as habits are all in one way or another culturally constructed, and often then perpetuated as norms. This is true of what we eat, how we farm and fish, the dominance of faiths be they founded on religious or other secular tenets as well as the technologies we deploy and the materials we use to meet our needs.

Material choice and use will undoubtedly continue to change, as they have throughout history. This is because substances to better meet needs have been discovered or developed, adverse consequences have manifested, or due to depletion or civil disruption in producing regions and associated cost hikes necessitating a search for alternatives or innovation of novel solutions.

Today though, there are more than eight billion mouths to feed, pairs of feet treading the land surface, bodies to house and feed, and brains to entertain, educate and employ. At the same time, we are facing massive resource depletion ranging from the availability of fresh water, erosion and degradation of soil, plummeting biodiversity and habitat quality, declining urban air quality, depletion of mined nutrients, rare-earth elements (naturally scarce substances with diverse uses in electrical, electronic and other applications and processes) as well as various metallic elements that are scarce in the Earth's crust and many more besides. Widening inequities between asset owners and disenfranchised communities deepen competition and the potential for conflict over dwindling resources. For example, there are growing demands, particularly from global regions with higher resource demands, for the conversion of tropical wetlands and other sensitive ecosystems to supply biofuels or food and cosmetic additives, leading local people to lose their supportive local landscapes and its associated biodiversity, the whole cycle of exploitation compounding already-alarming climate change trends. Although demands from the already-developed world place disproportionate pressure on resource

DOI: 10.1201/9781003637875-1

1

exploitation at a global scale, growth in population, per capita consumption and disparities in lifestyles within developing nations also amplify these trends. Resource choice, be that for food, fuel, construction or any other purposes, will increasingly be constrained as cumulative societal pressures bear down on declining natural biological and mined assets.

Hard biophysical limits will inevitably enforce change. We know that waiting to react to threats is generally, or almost invariably, disruptive and expensive, and often punitively so for businesses. Alternatively, we can choose instead to act upon what we know about how the world in which we live functions, determining proactively how to better integrate resource demands with the finite capacities of the natural processes and principles under which it operates. This can help us identify now the inevitabilities that can constitute a basis for informed innovation. We may thereby choose electively to recognise that tomorrow simply cannot be like today, but that needs will still exist and require novel means to meet them. This, in turn, can highlight necessary shifts in innovation, choice and profitable resource use as the pressures of natural forces tighten around us.

What is for certain is that we can benefit from thinking and acting differently. Either that or passively allow the inevitable to force us into picking up the pieces when current norms can no longer be sustained. Reorienting societal resource use patterns towards a sustainable destination is far from a trivial undertaking, as physical resources underpin all our activities. We know, though, that it is possible to rewire established norms as, throughout history, wholesale change has never dropped on us as a monolith: Japan's Tokugawa Shogunate, contemporary road and telephone infrastructures, the neoclassically rooted market nor the diversity of language systems encountered across the world fell out of the sky fully formed. Rather, every major transition throughout history – the Agricultural Revolution, the Industrial Revolution, the Information and Communications Technology Revolution and many more – has, in reality, been fomented as a cascade of smaller innovations led by change agents.

So, we know that rewiring societal norms is possible, but also that relying on retrospective responses to emerging and potential insoluble crises is no guarantor of averting disaster. The magnitude of threats now facing booming human demands dependent on dwindling planetary resources and support systems will require concerted application of knowledge, foresight and the building of consensus and intentionality. We also know that, as long as there are humans, there will always be needs to satisfy. Social and material 'satisfiers' will also therefore be required, no matter how much we dematerialise the ways we address some of them. Consequently, to meet many of our needs, we will always require technologies, products and physical materials. The materials we choose and innovate to continue to fulfil these diverse needs, and critically also their use patterns through societal value chains, will therefore also be a determinant of progress with addressing all strands of the complex sustainable development challenges now confronting us.

Who are the change agents rewriting societal norms and expectations about materials currently used, and who are the architects of innovations of novel materials into the future? Ultimately, every sector of society has roles to play, individually but particularly through collaboration, in progressing the journey towards sustainability and sustainable resource use. This understanding has to be contextualised within one of the objective realities of life on this planet: that no species can exist in isolation. Discoveries are revealing the close symbiotic interdependence of all multicellular organisms with a broader ecosystem of microorganisms. These holobiontic complexes, which we may formerly have regarded as discrete organisms, also interact with all other life forms within complex ecosystems exhibiting highly cooperative and mutually supporting connections in addition to more popularly understood competitive behaviours. Right up to the whole-planetary scale, tightly co-evolved ecosystems maintain energy, carbon, nutrient and other cycles that moderate the planetary atmosphere and in other ways act as an integrated 'superorganism'.

This is the ecosystem within which humanity evolved as a fully interdependent component. And, as we have progressed culturally and technologically – parallel strands that can be traced through serial innovations in material uses – the fact remains that all of our activities have a metabolism that is also fully interactive with and dependent upon planetary ecosystems at all scales. As we face increasingly daunting sustainability challenges, we often seek solutions at the institutional level: what companies do in terms of procurement, practices and products. But the reality is that no company is an island, all are integrated within value chains that conduct flows of energy, materials and human interactions. And, of course, these inherently connected business changes lie within a wider institutional ecosystem of regulators and legislators, markets, campaigners, knowledge-generators, consumers, municipalities and waste handlers. In other words, all players are part of a systemic and potentially symbiotic whole. Do we share common visions greater than the sum of these isolated component parts, providing us with a common purpose around which society can collaborate as a coherent whole towards the intentional achievement of sustainability, including the sustainable use of materials? If we fail to act in a cross-societally coherent manner to address widely accepted and at least rhetorically agreed on sustainability goals, fragmentation will condemn us and future generations to declining security and reduced opportunity. Working to achieve societal symbiosis guided by co-created visions can enable us to pull together to attain a better future for all.

Chapter 2, 'A Material Basis for the Meeting of Needs', addresses humanity as an element of the wider biosphere and how the discovery and innovation of materials has underpinned cultural evolution. Human needs, historically, now and into an uncertain future, require social and material solutions to satisfy them, emphasising the significance of material use for addressing all strands within the complex challenge of sustainable development. We stand at the leading edge of societal evolution, underpinned by a historic legacy that forms the habits of contemporary resource use. Only latterly have we become conscious of the breadth of social and ecological ramifications of the material use patterns that we have adopted. Developments in scientific understanding and analytical acuity have constantly changed the goalposts of what is considered safe, and they will continue to do so. This changing baseline of perceived safety has major ramifications for how we regulate chemicals, often formerly based on limited consideration of intrinsic chemical properties addressed in isolation from 'real-world' usage, exposure and risk. The well-known conception of risk as hazard multiplied by exposure is often overlooked, leading to misunderstandings about 'real world' safety but, as importantly, also the potential for different materials to support beneficial uses with appropriate risk-based scrutiny and management. Regulatory responses have traditionally been imposed reactively and with significant lag as acute concerns have arisen from resource use. But tomorrow cannot be a simple extension of today as booming human numbers and per capita resource use collide with the degraded and still-declining status of natural resources. New thinking about material use is essential to navigate by design towards a sustainable future in which needs are met in the safest and most efficient manner.

Chapter 3, 'The Good, the Bad and the Optimal', explores the ways that materials have been naively conceived as either intrinsically 'good' or 'bad' based on limited considerations divorced from contextualisation within the full societal life cycles of the products into which they are integrated. Whilst biologically based materials have roles to play in more sustainable approaches, over-simplistic assumptions that purported 'natural' materials are automatically sustainable are seriously undermined when their production, maintenance inputs during use and potential recyclability are addressed in a product life cycle context. Two take-home messages are that there are no such things as automatically sustainable materials and that some materials simplistically branded as 'bad' may be wholly consumed in production processes without risk, whilst others can imbue products with resilience conferring them with longevity and delivery of a high degree of societal value through efficiently addressing needs. And, if additive substances do not 'leak out' in use and are recoverable at end-of-life, this represents substantial continuing circular benefit delivery as a positive contribution to sustainability.

Chapter 4, 'Material Risks Over Whole Societal Product Life Cycles', expands on the importance of considering materials not merely in terms of their intrinsic properties in isolation, but in the context of their inclusion in the full societal life cycle of products. Examples of wider sustainability implications of material use are given across six life cycle stages: raw material extraction; material synthesis; material packaging and transport; compounding and converting; product use; and post-use. The need for a paradigmatic change in the assessment of material use to address the challenge of sustainable development is articulated, significantly including the importance of joining up risks at all stages of the life cycle into a comprehensive whole. Additive substances are given specific focus, including both those that are compounded with other materials to produce products and those introduced elsewhere across the whole life cycle, particularly including maintenance inputs during the product use phase that, for durable applications, may represent a substantial element of life cycle footprint. Refocusing thinking about the needs that material use serve is emphasised as the foundational framing under the 'Brundtland definition' of sustainable development. This needs-based focus, to which the world signed up as a consensual commitment in 1987, differs radically from the *de minimus* transposition of that ideal into common understanding and regulations based on incremental improvement and ecoefficiency, which fail to challenge underlying paradigms. These broader considerations inform opportunities to make progress towards a circular economy, the attainment of which requires the symbiotic collaboration of all sectors of society.

Chapter 5, 'Regulation as Enabler or Barrier to Sustainable Development', addresses a brief history of chemical regulation and the differing philosophical approaches underpinning the varying forms of chemical regulation now found across the world. Supporting chemical assessment tools are also briefly considered, noting that many are based on a bottom-up approach underpinning or driving improvement to reduce immediate adverse outcomes, but failing to stimulate progress in the 'Brundtland definition' spirit of driving innovation to better meet human needs both now and into the future. The ways in which regulatory approaches treat chemicals differently according to the purposes for which they are used are called into question. A concluding section asks if our legacy regulatory approach is fit for the purpose of stimulating sustainable progress. There is a clear need to reinterpret and revise historic Conventions, protocols and rules in the light of subsequent global changes and emergent priorities to ensure that they do not inhibit sustainable progress. Application of the precautionary principle is reassessed, reflecting on the tendency to prioritise immediate potential hazards over and above the greater challenge of innovation to meet needs safely and efficiently now and into the future.

Chapter 6, 'Innovation for a Very Different Future', revisits how tomorrow cannot be a simple extension of today due to the constraints of rising human numbers and demands dependent upon dwindling resources. This necessarily means that material use tomorrow will have to adapt to constraints within the socio-ecological system that supports ourselves and our activities. The chapter looks at sustainability-related trends to which we are adapting as well as shocks, or disruptive factors that were not predicted, that then ultimately constitute tomorrow's trends that need to be accommodated. Some of the tools commonly used to address chemical assessment are explored and evaluated against a set of principles germane to sustainable development, concluding that many address specific issues but fall short of embedding these wider principles. The chapter concludes with considerations about what constitutes strategic and profitable innovation, founded on thinking at the whole product life cycle scale and addressing all elements of sustainable development, to inform the types of material and use patterns that can most safely and efficiently contribute to the meeting of human needs in an inevitably different future. The role of the digital revolution is also considered in terms of its potential contributions to sustainable chemical innovation and use.

Chapter 7, 'Symbiosis for Sustainability', draws lessons from the symbiosis of organisms and ecosystems up to planetary scale, applying them to how the constituent sectors of society should also ideally operate interdependently as a coherent 'superorganism'. Progressive engagement with sustainable development issues by business and regulatory sectors leads to the potential for closer convergence around common aims, suggesting a new synergistic model to accelerate progress to shared goals. Convergence around common, co-created goals can also better harness the energies and investments of non-governmental and knowledge-providing sectors. Recognition of common, potentially distant goals can create 'pole stars' promoting cross-sectoral symbiosis, redirecting the energies of society from antagonism around current differences towards concerted progress towards collective vision. Examples of leadership integrating societal energies are found across business, regulation, voluntary and knowledge-providing sectors though, as yet, they represent 'baby steps' towards full symbiosis. Within this necessarily integrated, cross-sectoral system, a novel regulatory model is essential, built in partnership with businesses or business sectors around agreed sustainability goals. Unlike bottom-up regulatory approaches alone, which fail to challenge the paradigms underpinning current unsustainable material use, a novel symbiotic partnership between regulators and innovators can accelerate progress towards agreed aspirational sustainable goals, regulated through verification of audited achievement of targets leading towards longer-term goals. Shared visions with knowledge-providing and voluntary sectors can further accelerate progress towards sustainability.

Chapter 8, 'Realising Symbiotic Value Chains', extends thinking and practical steps towards the realisation of sustainable material value chains throughout society formed through coherence of intent, energies and investments of all sectors in society in pursuit of consensual end goals. These act as 'pole stars' for societal navigation towards sustainability that, though the final destination may be distant, offer strategic direction for innovation and the coherence of activities to inform stepwise progress on the journey. The journey to full symbiosis across society is represented as a 'pyramid model', from current antagonism at the four axes at its square base (rather like a boxing ring in which key players fight from their corners over current divisions) though with the apex of the pyramid representing coherence around common vision. This pyramid model is relevant to other societal challenges, with the use of materials as the means for their attainment.

Chapter 9, 'A View of the Journey from the Future', recognises that future generations will inherit and judge the consequences of actions, decisions and inaction. This chapter comprises a first-person reflection from the middle of the twenty-first century. A key narrative is one of corporate response to various 'shocks', jolting a material-producing business into a more far-sighted approach founded on collaboration across societal sectors around co-created visions of consensual sustainable end goals. Shifts in legislation and relationships with regulators, non-governmental organisations (NGOs) and others in the voluntary sector as well as the knowledge-providing community shaped by common visions are described. Still-unfolding progression towards increasing symbiosis across societal sectors represents a paradigmatic shift from old, antagonistic habits, instead working on the basis of a more liberating approach that harnesses cross-societal efforts and investments to make tangible collective progress towards consensual future goals that are beneficial to all.

Chapter 10, 'Accelerating Towards Sustainable Use of Materials', highlights the pressing need for and benefits of increasing symbiosis between societal sectors, integrating their best intents and energies around shared strategic goals to accelerate the pace of sustainable development. This builds upon scientific knowledge about supportive interdependencies across the socio-ecological system of which humanity is integral, and how learning from nature about cooperative strategies can be applied to harness energies between societal sectors. The social cohesion and cooperation necessary to implement a 'level playing field' approach as a material-blind framework for determining optimal material selection, use and innovation is discussed, reliant as it is on the fact that material flows through society have a significant social metabolism shaped by cross-societal collaboration. Harnessing of the energies and investments of all societal sectors through a socially symbiotic approach, to overcome contemporary dissipation through cross-sectoral antagonism, is essential to accelerate progress towards sustainability. The primary focus of this book is reprised: the quest for the sustainable use of materials, which forms the material foundation of a future of greater security and opportunity.

A Material Basis for the Meeting of Needs

2

Charles Darwin's famous evolutionary 'survival of the most fit' is commonly understood and promoted as taking the form of aggressive competition: Nature "*...red in tooth and claw*" as immortalised by Alfred, Lord Tennyson. The reality though is that fitness for survival is achieved with a high degree of cooperation both within and between species. Social structures are common to many species of animals, including both vertebrates and invertebrates, as well as between plants, microbes and bridging these divisions. This principle of fitness through cooperation is also highly germane to humans, including how we work together as societies and in operating the economy and the value chains and material uses that run through it.

2.1 A COOPERATIVE WORLD

A host of underappreciated cooperative strategies becomes increasingly evident both within and between species as we learn more about interactions between plants at the subsoil level. Close and often mutually essential supporting interactions between host flowering plants and pollinating animals, such as many types of insects, bats and other organisms, are just the most visible of these interdependencies between the plant and animal kingdoms that also, for example, include seed dispersal by fruit-eating birds, mammals, fish and insects. Symbiosis between plants and animals not only with each other but with microscopic organisms – bacteria, fungi, archaea, protozoans and many more types – is diverse and is also essential for survival. We are

DOI: 10.1201/9781003637875-2

learning that all macroscopic organisms are most likely holobionts, or in other words ecological units comprising assemblages of host and other generally microscopic species living within and/or around them creating and operating as single functional entities.

Ruminants are a graphic example of close biological interdependence between host organisms and symbiont microbiomes. These mammals ingest prodigious volumes of vegetation, yet few enzymes are coded by their own genome to metabolise complex plant polymers. Host animals rely instead almost wholly on enzymes produced by a diverse microbial assemblage retained within highly differentiated and specialised areas of the gut. The ruminant hosts rely entirely upon this microbial community to unlock nutritional value from what is eaten and also to synthesise important constituents such as short-chain fatty acids and vitamins. Climate change implications arising from significant methanogenesis by these internal microorganisms demonstrate a high degree of interdependence with all environmental media. The same level of metabolic dependency upon the microbiome is seen in a wide range of other animals such as wood-eating ants as well as fish that lack the enzymes necessary to break down lignin and cellulose. Baleen whales too ingest substantial volumes of krill and other planktonic crustaceans but depend entirely upon microbial processes in the gut to break down their chitin carapaces, liberating substances that the whales can then absorb.

Humans are in no way exempt from this interdependency. The endobiome of the human gut has profound implications for the health and functioning of the nervous, endocrine and immune systems. In fact, the number of symbiont microbial cells within the human gut, mouth, vagina, lungs and many other body spaces, as well as within the body organs and tissues, exceed those characterised by human DNA by a factor of three to one. Our internal microbes detoxify, digest, generate vitamins, fight pathogens, produce substantial proportions of the hormones regulating our bodies and interact closely with our nervous and immune systems. To this unseen legion, we owe our health and much of our mood.

Associations between macroscopic and microscopic organisms are not only deep; they are also integral. Whilst some amongst the microbial community are facultative or even opportunist, many are obligate symbionts with specificity not only to host organisms but also to specific regions of the gut, skin or other organs. The existence of distinct structures within animals and plants to accommodate them suggests that microbial and host organisms have co-evolved as unified holoorganisms, within which the endobiome plays significant roles from the embryonic stage and onwards throughout the whole life cycle.

We are deeply integrated with the ecosystems with which we co-evolved even at the subcellular level in terms of organelles that evolved as intracellular symbionts, organelles including mitochondria and chloroplasts possessing their own unique DNA. Environmental pollutants have been found to influence the constitution of the human microbiome, varying, for example, by geographical location, family structure and hence contact and exchange of microbes, as well as chemical exposure in the diet. Our endobiomes also vary with locality, for example with significant differences in internal microbial composition observed between people living in urban and rural locations.

At the whole-organism level, these symbiotic facets of humanity go well beyond the boundaries of simple physical competitiveness, albeit cumulatively conferring a competitive advantage to the holobiont, as we and other organisms have cooperative strategies at the community level. When one amongst us is ill, we seek to support and cure rather than to pursue competitive personal advantage from their death. We continue to be perceived as having valid contributions to make to society even after our biological duties are completed when our children become self-sufficient, and we have even invented the concept of pensions to resource us in our declining years. Optimally, we value the creative contributions of all in society, regardless of ability or disability, gender, race or other somatic differences. Discrimination serves only to limit our full social potential. Other animals have mutually supportive networks and intricate social structures, as do plants that cooperate via a subsoil 'wood wide web' of fungal mycelia both within and between species. However, it is a step up into cultural evolution that defines humanity.

From microbial and subcellular levels right up to the whole-planetary scale, ecosystems operate sustainably due to tightly linked interdependencies that evolved over the 3.85-billion-year history of life on Earth. Life interacts in turn with all environmental media and their physical and chemical constituents as a cyclic and integrated whole, comprising a planetary-scale holobiont, or 'superorganism' as famously conceptualised under the Gaia hypothesis.

The purpose of this deeper consideration of recently realised, yet-to-be-discovered or even wholly unsuspected biological impacts linking organisms to chemical constituents in the environment and in our diets is not to raise alarm. Rather, it is to highlight that our knowledge is very far from complete and is constantly evolving. For this reason, substances we've considered safe yesterday have subsequently emerged as of concern on the discovery of more subtle and often chronic consequences, such as endocrine disruption, teratogenicity and other impacts. This trend of new discoveries will inevitably continue as we learn more about biological effects that may have profound implications for continued healthy functioning not only of human cells but also of the organisms with which we share this world and the environmental processes that they perform. It would be folly to think that our knowledge in this regard is complete.

2.2 THE MATERIAL BASIS OF CULTURAL EVOLUTION

Humans go beyond simple competition through the evolution of a collective consciousness, sharing of knowledge leading to cascading accumulation of discoveries, belief systems and accepted codes of conduct. Arguably, we have called time on physical evolution by using our snowballing intellectual capacities to manage women and the unborn through difficult pregnancies or even to promote in vitro fertilisation for those unable to conceive. We also innovate increasingly sophisticated medical interventions to prevent sick people from being edited out of the gene pool. But it is in the virtual world that we humans forge ahead. Millennia-old philosophies and resource use practises inform us still today, and every day we continue to extend them as humanity's collective knowledge expands and adds to the cumulative pool of consciousness.

Without getting too philosophical, aspects of human identity are coded into nucleic acids, but the expression of humanity requires incarnation of that digital genetic information into bodily form and functions. That process requires material exchanges with the surrounding environment to meet changing needs from conception to death.

Human needs have attracted study and definitions by many authors. They have also featured in initiatives such as the 30 Articles of the UN Declaration on Human Rights[1] as well as the UN Sustainable Development Goals.[2] A framework by Abram Maslow represented human needs hierarchically, ranging from basic physiological needs (such as for food, sleep, sex and shelter) forming foundations for higher tiers of needs including those relating to safety, belonging and esteem, all supporting the potential for 'self-actualisation' (achieving individual potential').[3,4] Manfred Max-Neef rejected the hierarchical approach but helpfully elaborated how all interlocking needs have their satisfiers – the material and immaterial things necessary to fulfil the needs – but also a plethora of 'false satisfiers' as observed in addictions to drink, drugs, food, sex, toxic relationships and consumerism that, though offering short-term relief or anaesthesia, ultimately fail to satisfy deeper needs such as for love, companionship and esteem.[5]

Satisfaction of needs forms the basis for humanity's intimate and evolving relationship with physical materials, driving the exploitation of many substances to fulfil them. Plumbing, ventilation, food supply, communication and transport systems as well as books, libraries, schools, homes, offices and communal spaces in urban areas constitute physical infrastructure enabling the meeting of a broad spectrum of biophysical, social, creative and spiritual needs. The innovative means we have deployed to fulfil our increasingly complex needs have shaped pathways of human development and culture, including taboos as well as the exploitation and innovation of different materials.

Consequently, cultural evolution leaves a trace of shifting patterns of material exploitation or innovation. The use of water and wood is traced by the earliest artefacts of human history, through the well-known Stone, Bronze and Iron Ages, and onward to the steel and coal of the Industrial Revolution and the semiconductors underpinning the massive connectivity and data handling of the ICT (Information and Communications Technology) Revolution. All have sequentially opened up new capabilities, enabling the meeting of needs – though sometimes also regrettably with a collateral proliferation of false satisfiers – in more efficient and complex ways.

2.3 THE WAY THINGS ARE

In our everyday lives, we may not recognise that we stand on a roadway of constant material and cultural evolution. We have inherited norms and assumptions, hard-won and hard-wired over extended histories. The things we reach out for to meet our needs, including the materials from which they are constructed, may have a long legacy, as well as frequently a deeply entrenched set of vested interests. As the old joke goes, a driver stopped in the countryside to ask a local for directions to a particular city but, after some prevarication and lots of changes of mind, the local eventually told the driver that, ideally, he shouldn't start from here. On the roadway to the future, we stand where we are today, a place that is far from optimal from a sustainability perspective. Our current location and habits are constructs of assumptions, norms and resource use patterns with a long history in terms of the materials we deploy.

Have we seriously thought about the most appropriate and sustainable use of materials in the products we reach for to meet our needs in an inevitably changed future? Or have we simply accepted the things that have shaped today's inherited norms? What then is 'normal'? There are many definitions of the word. Most revolve around the concept of conformity to standards or conditions that are typical or expected. The word itself is derived from the Latin 'normālis', describing something made with a carpenter's square (which is the reason that 'normal' is also used to define something perpendicular to a line or surface). The basic principle at play though is consistency with a rule.

Many things in this world are 'normal'. It is just the way the world is, albeit there is a high degree of cultural specificity depending whereabouts on the planet we live. In the UK, India, Japan and Australia, we drive on the left-hand carriageway of the road and cars are constructed and marketed with that orientation, whereas across continental Europe and the US, the 'normal' rules of the road are that one drives on the right-hand carriageway and cars and road furniture are built for that purpose. Toilet etiquette varies from the western world's use of toilet paper to cleansing with the hand in much of the South Asian world, and particularly the left hand only in many places. In some sectors of Western society, there is strict etiquette about the sequence in which cutlery is used, whilst chopsticks are the norm in many East Asian countries and, in much of the less-developed world, people generally eat with their hands and then also generally only the right hand. A 'thumbs up' is polite in many countries but deeply offensive in others, as indeed is burping at the dinner table. The traveller is well advised to learn beforehand the norms of the places they visit. And, of course, normality changes over time as well as space, as evidenced by widespread practices that were prevalent then fell out of favour and were ultimately outlawed, at least in some locations, such as smoking in public places, failing to wear a seat belt whilst driving, or using lead additives in paint, road fuel, plastics or even, in Elizabethan England, as a cosmetic to lighten skin tone.

All of which might loosen our assumptions about what is normal in an objective sense. The normal ways things are done is merely a facet of what has come to be adopted and expected in our niche in space and time, and for which we now have sunk economic models. We rarely pause to ask why things are the way they are. And this, of course, extends to the foods we eat, the drugs we administer and the materials we deploy in the diversity of applications and products supporting our daily lives.

2.4 SOCIAL AND ECOLOGICAL LINKAGES

A biophysical reality of the planetary system within which we co-evolved is that everything is intricately interconnected. All organisms, humans included, eat, excrete and interact with the supporting planetary ecosystem in myriad ways. Also, all of our broader economic and other activities have a metabolism that interacts intimately and unavoidably with the supportive biosphere.

When we think of materials and their uses, we need to think not only of their utility but also about where they've come from, their interactions with the processes of the wider world in how we use them, and where they go when we've finished with them. This equally unavoidably also includes the intimate linkages that we have with everyone else who shares the biosphere and its supportive processes. If we overharvest, we not only damage the potential of ecosystems to sustain us into the future but we also create inequities for their co-dependents. We live in a deeply integrated socio-ecological system, not to mention a densely populated world, wherein every action interacts with the biosphere including all co-dependent species, including humans. This is maybe not a perspective that we've been brought up to think about in terms of the materials in objects we use on a daily basis, but it is nevertheless a fixed biophysical reality.

Awareness of the finite limits of socio-ecological systems, and the magnitude of the pressures of growing human numbers and increasing technological reach upon them, has underpinned the progressive development of recognition of the problems, both familiar and emergent, facing the world. This awareness in turn has framed the need to reorient society onto a sustainable pathway of development.

The story of how looming environmental crises have disrupted societal comfortability and enforced growing concerns about the depth of unsustainability of the world today is told elsewhere. Suffice it to say that the magnitude of these cumulative pressures is potentially now existential; the triad of crises of climate, biodiversity and pollution are increasingly entering the mainstream of societal discourse if, as yet, failing to precipitate proportionate action. We can add to these pressures pervasive and increasing water stress as well as food insecurity. Further dire threats include current and potential civil instability consequent from resource depletion and greater competition for what remains, exacerbated by inequities in society that see the rich growing richer at the expense of the poor growing increasingly resource-poor as the common wealth of supportive ecosystems continues to decline and degrade. We are already handing a greatly impoverished inheritance to future generations, compromising their abilities to meet their needs contrary to the rhetoric of intergovernmental proclamations about a commitment to sustainable development.

Clearly, there is a collision between human demands and environmental capacity. It is already widely evident with varying severity throughout the world, for example in the exhaustion and erosion of soils as well as water stress evident in many regions, not to mention the depletion of natural resources and the accumulation of pollutants with feedback into aspects of human health, economic activities and wider wellbeing. Radical revision of the course of development respecting the interdependence of the whole integrated socio-ecological system is clearly essential.

One of the consequences of the substantial scale of human occupation of and demands on planetary resources, and the implications of the waste materials emitted into the biosphere, is that socio-economic processes substantially augment ecological processes, with the socio-metabolic regime of contemporary industrial society now comprising principal drivers of today's mounting sustainability challenges.[6]

2.5 SUBSTANCES OF CONCERN

Concerns about the presence and possible accumulation of heavy metals, pesticides and other persistent organic and inorganic substances, both in the environment and in human cells, have rightly been a prominent feature of the early environmental movement and regulatory responses. Various national, regional and global statutory instruments have been implemented to address human health and environmental risks associated with the accumulation of heavy metals. As one global-scale example, an international treaty known as the *Minamata Convention on Mercury* was signed in October 2013 by nearly 140 countries aiming to reduce anthropogenic emissions and releases of mercury and mercury compounds to protect the human health and the environment, addressing mining, export and import, safe storage and disposal.

Non-governmental organisation (NGO) activism about chemical warfare agents, in particular organophosphate and organochlorine substances that have common chemistry as both nerve agents and pesticides, was also a significant strand of the early environmental movement, paralleling the influence of Rachel Carson's seminal 1962 book *Silent Spring* (discussed later in this chapter). Pernicious threats associated with the spread, dispersal and potential accumulation of persistent organic substances led to the signing of the *Stockholm Convention on Persistent Organic Pollutants* in May 2001. The aim of the Stockholm Convention was to eliminate or restrict the production and use of persistent organic pollutants (POPs), defined as "*...chemical substances that persist in the environment, bio-accumulate through the food web, and pose a risk of causing adverse effects to human health and the environment*", following negotiations after a United Nations Environment Programme (UNEP) called for global action in 1995.[7]

Radioactive materials, too, have been a focus of concern, particularly in light of the atmospheric detonation of nuclear weapons leading to their global dispersal and rising concentration. The testing of nuclear weapons began on the morning of 16 July 1945 at a desert test site in Alamogordo, New Mexico, when the US exploded its first atomic bomb. Nuclear bombs were dropped on Hiroshima and Nagasaki, respectively, on the 6th and 9th of August that year. Between 1945 and 1996, over 2,000 nuclear explosions occurred worldwide, of which over 500 were detonated in the atmosphere as the UK, the Soviet Union, China, France and India joined the US as states with nuclear capabilities. Pakistan and the Democratic People's Republic of Korea (DPRK: North Korea) were shortly to join the 'nuclear club', with others following. Atmospheric testing was banned under the 1963 Partial Test Ban Treaty signed on 5th August by the US, the Soviet Union and the UK, largely in response to grave concerns expressed by the international community regarding radioactive fallout. France and China declined to join the Treaty, though conducted their last atmospheric tests, respectively, in 1974 and 1980. The *Comprehensive Nuclear-Test-Ban Treaty* was subsequently opened for signature in 1996 with the intent of banning nuclear testing everywhere on the planet, including the land surface, atmosphere, underwater and underground. By August 2024, 187 countries had signed and 178 ratified the Treaty. However, of the 44 states listed in the Treaty as then possessing nuclear capabilities, nine were missing either as signatories or having failed to ratify (China, DPRK, Egypt, India, Iran, Israel, Pakistan, the Russian Federation and the US) meaning that, technically, the Treaty has yet to enter into force. Nonetheless, this demonstrates an advanced level of international cooperation around addressing risks associated with the release and accumulation of hazardous materials, extending to tighter controls of non-weapon uses of these dangerous substances.

Other transboundary pollutants have come under increasing scrutiny, particularly aerial pollutants that do not respect national jurisdictions leading to regional or pan-global circulation and deposition. These substances can also be emitted from sources in international spaces, such as shipping and aviation. Control of these pollutants calls for regional and global cooperation, including acknowledgement of sources in one territory that may manifest as problematic in another. A number of these potential pollutants are the subject of the UNECE *Convention on Long-Range Transboundary Air Pollution* (LRTAP) signed by 32 countries in a pan-European region in 1979. Agreement to phase out substances implicated in the depletion of stratospheric ozone culminated in the signing of an intergovernmental treaty in September 1987, known as the *Montreal Protocol on Substances That Deplete the Ozone Layer* (the 'Montreal Protocol'), with periodic subsequent revisions. Its projected success led former UN Secretary-General Kofi Annan to declare that "...*perhaps the single most successful international agreement to date has been the Montreal Protocol*".[8]

Climate change emerged during the latter quarter of the twentieth century as an issue of major disruptive, potentially existential threat. Accumulation of carbon dioxide in the atmosphere gives particular cause for concern, with attention also being paid to other climate-active gases such as methane and nitrous oxide. Global monitoring and policy recommendations pertaining to international cooperation to address climate change flow from the *United Nations Framework Convention on Climate Change* (UNFCCC) through its regular reports. Undoubtedly, substances contributing to the instability of the climate are a matter of major transboundary pollutant concern at the global scale. Curtailment of climate active gases is consequently emerging into policy instruments around the world, though action remains far from proportionate to the threats it poses with concentrations of carbon dioxide in the atmosphere still rising despite an abundance of political rhetoric at international and national levels.

It would be a mistake to assume that all cases of pollution are related to synthetic substances. In addition to the mining of metals, sequestered phosphorus and fossil carbon from the lithosphere, other substances present in the biosphere are converted into novel forms. A significant pollutant in that respect arises from the conversion of atmospheric nitrogen from its largely inert molecular form (N_2 comprising 79% of the air we breathe) into bioavailable forms that contribute to the eutrophication of water and soils. Natural processes, particularly lightening and biological transformations performed by microorganisms, capture nitrogen from the atmosphere and convert it into reduced forms that are then available to plants and other organisms, though at low concentrations that limit natural primary productivity. However, industrial means for fixing nitrogen from the atmosphere have increased exponentially since their innovation in the 1940s, with a prediction that human-mediated fixation of nitrogen will exceed that fixed by microbial processes by 2030.[9] This is inherently problematic as eutrophication with nitrogen exerts profound changes upon ecosystem structure, function and resilience globally.[10]

What all these substances have in common is that they are either concentrated or produced by society with the risk of causing deleterious impacts, particularly where they are prone to accumulate leading to thresholds that may be known or unknown, beyond which harm to human and environmental health and natural processes may occur.

2.6 MATERIAL SAFETY

Safety matters hugely. Prejudices about some materials and the uncritical acceptance of norms aside, we generally look to science for guidance.

One of the phrases in common use that tends to annoy me and many other scientists when we hear it used uncritically in the media or in general discourse is "*Scientists say that...*" Sometimes, this is a genuine attempt to reach out for some solidity to underpin a view. However, often it is lazily picking up isolated facts to support a particular argument rather than an attempt to ask critically what differing scientific perspectives suggest when viewed in the round. Beneath this is a common misunderstanding of the nature of science. Whilst, in part, science relates to the search, establishment, accrual and use of facts, it is more truly a deeper quest for understanding. It is questing, testing and, at any one time, an amalgamation of 'best so far' consensus better to guide society. It is certainly not infallible, and consensus can and does change with new insights. The world is not flat, the Earth is not the centre of the universe, 'germs' are not exclusively bad things, the human appendix is not vestigial, atoms are not the smallest divisible constituents of matter, the colour red does not make bulls angry, and humans have tens more senses than the formerly assumed five.

2.6.1 Are We Certain It's Safe?

When it comes to the assessment of the safety of materials, there is unavoidably a substantial degree of uncertainty. Looking back in history, we can see that wonderful innovations, like asbestos enhancing fire safety and lead anti-knocking agents for greater fuel efficiency, turned out to be rather less beneficial than initially anticipated. Further examples are to be found in the legacy of drugs – thalidomide and also both prescribed and illegal opioid use in the US are pertinent – that had intended beneficial consequences but unforeseen negative implications. Add to this a sequence of different pesticides used in agriculture that promised great things but were later revealed to be pernicious poisons, leading to a carousel of substance substitutions from arsenic to methyl bromide, organophosphates, organochlorines, synthetic pyrethroids and today's neonicotinoids with all of their now apparent disastrous impacts on insect populations, wider environmental damage and human health concerns.

This is not to say that we should distrust science, but rather that we should better understand it. Science gives us guidance, but this is based only on what we know today. Examples of promotion of diesel road fuel for carbon-saving naive to generation of fine particulate emissions, use of radium to illuminate watch dials, hydrogenation of oils and their inclusion as transfats in food, social uptake and former advice on beneficial use of tobacco, unforeseen consequences arising from the use of various pharmaceuticals and chemical additives, and primary materials of all types such as lead in plumbing should inform us that today's 'good' substances may be tomorrow's legacy should we discover that there are wider issues we have not yet foreseen. We essentially make the best judgements we can within the limited span of contemporary knowledge.

2.6.2 A Changing Baseline of Safety

In my time in government, awareness of the endocrine-disrupting impacts of chemicals, of which there are many and diverse types, began to emerge into public and regulatory consciousness. These were found to generate biological impacts, largely formerly unanticipated, sometimes at several orders of magnitude lower concentrations than prior 'safe' standards based on simpler ecotoxicological parameters. It has taken many years for the regulatory environment to grapple with the problem of endocrine-disrupting substances, let alone to manage them effectively.

Add to this the ever-evolving acuity of analytical equipment. As a PhD student, I worked with heavy metals because the novel atomic absorption technology of the day allowed us to analyse them in an affordable way at finer concentrations than formerly. Yet, a subsequent generation of research students left behind metals to work on organic substances, substantially due to the emergence of high-performance gas and liquid chromatography that made their detection possible at a reasonable cost. Today, laboratories are equipped with DNA analytical machinery exploring still different aspects of the impacts of chemicals and other factors at genetic and epigenetic levels. In many ways, the acuity of emerging analytical technologies drove our concerns about different substances, rather than concerns about substances necessarily driving a quest for greater analytical acuity.

And then there's the behaviour of substances themselves. We know, for example, that radioactivity has no safe threshold. We are learning that other substances also have impacts that cannot be reliably related to 'safe' toxic thresholds, but that have a continuum of effects including, for example, many endocrine-disrupting, carcinogenic and reprotoxic substances. This is inconsistent with the legacy regulatory approach of setting 'safe' environmental quality and health-related standards. Furthermore, as our analysis of DNA and other attributes of life grow in acuity, we are finding ever more subtle impacts that were formerly unforeseen. Yesterday's 'safe' limits are exposed as being founded on what we knew previously, subsequent discoveries exposing our naivety.

And then we have the behaviour of substances in the environment to add further complexity. For example, the difference between chemical behaviour in vitro and in vivo may differ significantly due to factors such as chelation, methylation, variable pH leading to speciation, or wider changes in the properties of different substances in complex natural environments or within the human body. A particular substance may have a synergistic effect with other pollutants as indeed other natural substances within the environment. This can create unexpected threats, compounded further by potential breakdown into metabolites that may in turn have further concerning implications.

2.6.3 Safety, the Microbiome and Ecosystem Processes

It is inevitable that further impacts will be discovered as a consequence of the release of chemicals into the environment and into the human body due to the factors addressed above. However, a further, potentially substantial complicating factor virtually guaranteeing that more subtle effects will become evident over time is the growing awareness that all multicellular organisms, at least all that have been studied, are in fact holobionts. These macroorganisms are fully interdependent with symbiotic microbial organisms such as bacteria, archaea, fungi, spirochetes and other protozoa, viruses and many more, all of which make significant, often essential functional contributions to the workings of the whole organism.[11] Rooted plants, for example, cannot function without their attendant rhizobiome (the microbial community closely associated with their roots) to which plants excrete typically 15%–45% of their net production of carbohydrates in exchange for microbial processes that enable access to soil water, minerals and other constituents as well as combatting potential pathogens.

As already noted, the human body comprises three times as many non-human cells as those containing human DNA. Looking just at the endobiome of the gut, these microbes are deeply interdependent with the human host, playing significant and vital roles in digestion and nutrition including synthesising vitamins and breaking down larger molecules into moieties that the gut can absorb. They detoxify potentially harmful substances and suppress potential pathogens. The inner microbial community also has deep interactions with the human nervous, immune and endocrine systems. A diversity of substances released by the microbiome has been found to exert significant influence on the central nervous system, regulating brain chemistry and metabolic processes with implications for stress response, anxiety, memory function and mental health.[12,13] The enteric nervous system, a mesh-like system of nerves governing the functioning of the gastrointestinal system sometimes referred to as a 'second brain', contains around 500 million neurons (approximately five times as many as in the spinal cord) and operates in close symbiosis with the gut microbiome with a wide range of direct impacts on the whole body. Enteric microbes produce neurotransmitters directly, also influencing the production of numerous hormones synthesised by enteroendocrine cells located in the mucosal lining of the intestine. This complex of enteric nerves under the influence of microbial symbionts produces more than 90% of the whole body's serotonin (a neurotransmitter hormone touching on a wide range of functions including mood, cognition, reward, learning and memory amongst other physiological processes) and approximately 50% of dopamine (a neurotransmitter connected with reward-motivational behaviours, motor control and the release of various other hormones).[14,15] Through this profound symbiosis, the microbiome exerts a diversity of impacts including the social and behavioural responses of, and epigenetic expression in, host organisms. Chemical impacts upon this complex of functionally important and irreplaceable community of microorganisms in both humans and other multicellular organisms are largely under-researched but will be likely in future to yield new insights about chemical impacts, bringing into question former assumptions about safety based on simple single-species ecotoxicological analysis.

These microbiome impacts relate to both synthetic and natural substances. For example, the consequences of cavalier use and release of antibiotics may affect environmental pathways, in addition to contributing to the rise of resistant microbes creating a looming threat to human health. Currently unknown impacts, both subtle and more severe, on a host of other organisms with roles in the integrity and functioning of holobiontic organisms and wider ecosystem processes may represent further potential threats. Whilst we have to remember that potential hazard does not equate automatically to risk unless there is a clear exposure pathway, avoidance of the most damaging known substances remains a wise strategy, as does the circular recapture and reuse of spent materials averting their systematic accumulation in natural systems potentially triggering consequences as unforeseen thresholds are breached.

Further complexities remain to be discovered in terms of the impacts of substances on ecosystem processes. In the wider ecosystem, a complex of organisms, particularly the microbial community, plays vital roles in geochemical cycling and the wider functioning of the whole system and the ecosystem services that it provides. Many of the functions generating the ecosystem services upon which humanity depends or otherwise enjoys are far from fully understood, and the role of biodiversity in their generation is complex, multi-layered and requires a great deal more research.[16] New knowledge about chemical impacts on these supportive ecosystem processes is likely to emerge in the years ahead, further challenging perceptions of safety and requiring us to revise our approaches.

The intent of this discussion of science and safety is not to shatter faith in science and regulation. Rather, it is to shine a light on the fact that we're doing the best we can at any one time. The world we live in is complex and we are constantly learning more about it, certainly as we move from an overly narrow focus on single concerns addressed in isolation. So, branding of 'safety' is, in reality, a momentary and evolving thing, bounded by context and with inherent associated uncertainty. It is therefore essential that we adopt an adaptive approach to determine what is safe and we apply that to how we use materials to meet our needs in the safest and most efficient manner that we know at any one time. It is also vital that we integrate this adaptive and, above all, systemic approach into regulatory processes.

2.7 REGULATORY RESPONSES TO GROWING AWARENESS

A wide range of emerging problems have progressively jolted society out of former comfortabilities. The consequent emergence and progressive development of regulatory controls on chemicals in the environment and potentially affecting human health will be addressed in further detail later in this book. However, a brief overview is given here in relation to how differing perceptions of threats posed by chemicals have shaped regulatory and control mechanisms.

2.7.1 Waking Up to the Threat of Persistence and Bioaccumulation

The incidence and naming of 'Minamata disease' in 1956 in the environs of Minamata Bay in Japan's Kumamoto Prefecture served as a particularly shocking wake-up call about the adverse implications of accumulation of persistent pollutants in the environment. A wide range of neurological damage, disabilities and developmental harm resulted from chronic mercury poisoning due to long-running releases of methylmercury in industrial wastewater from a chemical factory discharging into the bay between 1932 and 1968. Mercury was subsequently bioaccumulated and biomagnified by shellfish and finfish in both Minamata Bay and the surrounding Shiranui Sea and was, in turn, consumed by an adjacent human population highly dependent upon locally harvested seafood as a primary protein source. Deaths and disability of humans and animals continued for a period of 36 years. Corporate and government response was slow, with official recognition of the disease and compensation initiated as late as March 2001, and final settlement of victims in March 2010. Environmental protests relating to the disease are claimed to have played a significant role in the democratisation of Japan.[17] What is certain is that the horrors of Minamata disease presented the world with a wake-up call about the consequences of cavalier discharge of persistent substances, and the devastating health consequences and liabilities that could ensue as they accumulated and entered human and animal food chains.

Another significant stimulant of broader awareness about the consequences of unquestioning and cavalier release of persistent and bioaccumulative substances into the open environment occurred in 1962 with the publication of the seminal book *Silent Spring* by the American marine biologist, author and conservationist Rachel Carson. *Silent Spring* was the most globally impactful of Carson's trilogy of books, credited with not only advancing marine conservation but also initiating the global environmental movement. A stark warning was presented in the opening words of *Silent Spring*: "*No witchcraft, no enemy action had silenced the rebirth of new life in this stricken world. The people had done it themselves*". The evocative title of the book too conjured a dystopian image of a spring robbed of life-affirming birdsong. This demonstrated Carson's skills not merely in synthesising a growing scientific evidence base concerning the persistence and bioaccumulation of pesticides even remotely from where they had been applied but also in expressing it in emotive and intuitive terms that drew the attention and concern of media and the wider public. *Silent Spring* is credited with driving presidential initiatives in the US that led to dichlorodiphenyltrichloroethane (DDT) being banned within a decade, albeit Carson had implicated a far wider range of pesticide substances with similar pernicious implications. The book was also to influence awareness and concern at a wider global scale regarding uncontrolled application and the release of persistent substances into the open environment.

The 1970s constituted a particularly significant decade in the rise of global environmental awareness. It saw the consolidation of the 'environment movement' as we regard it today, and the beginning of substantive regulatory controls on many aspects of the environment including on chemicals. The 1970s was also a decade that saw the initiation of thinking about environmental economics and novel approaches to environmental management, including the control and release of chemical substances.[18] 1972 constituted a milestone year in terms of broader global awareness and action regarding human interdependence with planetary ecosystems and the threats represented by the parlous decline in environmental quality and carrying capacity. The United Nations' 'Stockholm Conference' (the UN Conference on the Human Environment[19]) in 1972 drew the global community together around these emerging concerns. 1972 also saw instigation of the UNEP and was also the year that the Club of Rome's groundbreaking *Limits to Growth* report[20] was published.

2.7.2 Recognition of the Linkages between Ecology, Society and Economy

Precipitous decline in biodiversity and habitats, both locally and at a global scale, also began to frame public concern about the need for a different approach to nature conservation. A key global initiative embodying this change in focus was the establishment of the Ramsar Convention, signed by an intergovernmental gathering in the Iranian city of Ramsar in 1971. The Ramsar Convention, known as 'The Convention on Wetlands', recognised that wetlands could not be conserved by a 'fortress conservation' model excluding human activity. Rather, wetlands were recognised as a primary resource for the wellbeing, activities and cultures of people, and that the conservation of these multifunctional ecosystems needed to be addressed through the sympathetic interaction of social, economic and ecological elements of the systemic whole. This led to definitions under the Convention of 'Wise Use' and protection of the 'Natural Character' of wetlands, upon which their continuity depended. A major stimulus behind instigation of the Ramsar Convention was the role that wetlands distributed across the world play in global flyways of migratory birds that are dependent on these systems but were, and still are, observed to be in rapid decline. The systemically connected triad of social, economic and ecological aspects of inherently integrated socio-ecological systems has progressed as an underpinning model for future conceptualisation and policy development about a sustainable pathway of development.

Concerns about the declining condition of the natural world and its ramification for human wellbeing led to the development and publication of the *World Conservation Strategy* in 1980 by a consortium of international environmental NGOs.[21] The *World Conservation Strategy* was the first international document on the global state of living resources and the need for their conservation that had been produced with combined inputs from governments, non-governmental organisations and a range of other experts. It was significant for arguing for a sustainable path of development founded on the protection of ecological processes, life-support systems, genetic diversity and sustainable utilisation of species and ecosystems, mirroring the approach of the Ramsar Convention but extending to all habitats and spheres of human activity impinging upon them. Further elaboration by intergovernmental collaboration advancing the concept of sustainable development – advancing the concept of the three interlinked constituents of social, ecological and economic progress – occurred under the World Commission on Environment and Development (the 'Brundtland Commission' named after its chair) leading to the influential 1987 report *Our Common Future,*[22] more widely known as the 'Brundtland Report'. This highly influential 1987 report brought the principle and language of sustainable development into wider media and political discourse, along with consensus about the 'Brundtland definition' of "...*development that meets the needs of the present without compromising the ability of future generations to meet their own needs*".

Successive global gatherings relating to the interactive relationship between humanity and the environment and the pursuit of sustainable development have occurred since that time. Significantly, this included the United Nations 'Earth Summit' held in June 1992 in Rio de Janeiro, Brazil. This has led onwards to a series of decadal intergovernmental gatherings, albeit consensus commitments have subsequently seemed to comprise increasingly unpersuasive compromises.

2.7.3 Emerging and Contrasting Approaches to Chemical Regulation

Concerns about chemical safety led to the enactment of environmental legislation and the founding of regulatory bodies, particularly in the 1970s. Chemical regulation has evolved at varying paces and with differential rigour around the world. The way that risks associated with chemicals are addressed also varies significantly between different global regions.

In the US, the Environmental Protection Agency (EPA) was established in 1970 as an authority charged with aspects of environmental protection. The EPA was initially charged with the implementation of the US Clean Air Act (1970), and with developing and implementing consistent national guidelines and enforcement regarding environmental regulation replacing a diversity of often conflicting and generally ineffective laws pertaining to environmental protection enacted at state or community level. The remit of the EPA expanded over time to include pollution control programmes such as monitoring water and setting discharge standards for industry and municipal effluents under the Clean Water Act 1972, authorisation and regulation of pesticides, and control and clean-up of waste sites. In the twenty-first century, the remit extended to address climate change. In terms of controls on the use of chemicals, substances that were in use before the EPA became the competent authority in 1974 have only slowly been addressed, with many pre-existing substances of potential concern authorised under a 'grandparenting' approach rather than having received systemic and objective scrutiny.

In Europe today, the EU REACH (Registration, Evaluation, Authorisation and Restriction of Chemicals) regulations claim to take a risk-based approach though, in reality, the criteria used for assessment are based on the 'intrinsic properties' of substances or, in other words, consideration of hazard outside of the context of their use and the degree of exposure that this might entail. Only once substances are identified as candidates for restriction on the basis of their intrinsic properties are wider risk considerations brought to bear under the REACH process despite hazard, risk and safety not matching closely if 'real world' exposure from use is excluded from consideration.

2.7.4 Hazard versus Risk

Consideration of potential hazard alone in the absence of 'real life' context can lead to unhelpful judgments about actual risks posed by substances. If embedded in regulatory approaches, this can lead to unhelpfully restrictive policies compromising the potential utility of substances in meeting societal needs.

Water is commonly considered a safe material and one that is crucial for life, yet an estimated 236,000 annual drowning deaths occur worldwide constituting the third leading cause of unintentional injury deaths.[23] Another safe and essential substance, oxygen, can become an instantly fatal neurotoxin when critical partial pressures are exceeded, this risk constituting a key part of the training of scuba divers using oxygen-enriched Nitrox gas mixes. Basic scuba training also warns that the 'inert' gas nitrogen can have significant narcotic effects affecting judgement below critical depths. Let us also not forget that too much inert material can also be lethal, with the American state of Alabama joining the states of Oklahoma and Mississippi in 2018 in permitting nitrogen hypoxia as a means of execution. These examples of everyday 'safe' and 'inert' or 'benign' substances, which can be either fatal or create substantial risk in different contexts, are illustrative of the potential for poor decision-making and regulation of substances when divorced from the wider context of risk.

The same distorted perspective of actual risk can be associated with chemicals with more hazardous inherent properties. For many chemical precursors in industrial processes, tightly controlled manufacturing controls may see them wholly consumed with no exposure and hence no uncontrollable risk. The same observation is true of substances immobilised in compounded materials and products, or inherently toxic metals such as copper in widespread use in plumbing, which may then be recovered and recycled post-use.

Regrettably, narrowly hazard-based regulation – easier and cheaper to develop and enforce than an approach founded on actual risk – can inhibit the use of chemicals that may beneficially support human needs. The regulatory environment remains significantly adrift today from one that is truly risk-based and related to the best ways to serve the sustainable goal of meeting needs.

One of the many ramifications of common presumptions and associated regulatory approaches based on simplistic hazard criteria alone, divorced from the context of their use in the whole societal life cycles of products, is a pervasive yet wholly misplaced assumption that materials may be inherently and automatically 'good' or 'bad'. This is such an important topic that Chapter 3 is dedicated to it.

I should at this point issue a 'spoiler alert': the notion of an inherently sustainable material is a fantasy. The sustainability of materials must be addressed in terms of their use across the whole societal life cycles of the products in which they are integrated, along with wider dimensions of potential for chemical contamination, resource depletion and the meeting of needs in the safest and most efficient manner. The sustainable use of materials, as distinct from inherent material properties in isolation, is a theme to which we will return throughout this book. We will also relate this to the need for a 'material blind' approach to addressing the most sustainable choice and use of all types of material in different contexts and applications on a systemically informed 'level playing field' of principles.

2.8 THE FUTURE IS A DIFFERENT COUNTRY

The saying *"The future is a different country"* is apposite to many situations in our dynamic world. This saying is equally relevant to the materials that society uses to meet its needs.

The happenchance of discovering how materials in the biosphere can be used to address our needs, subsequent exploitation of substances found in the lithosphere and synthesis of a bewildering array of substances entirely new to planetary history has enabled us to meet needs in increasingly innovative and efficient ways. However, regarded with current insight through the telescope of history, we can see that we have too often applied our material discoveries without awareness of their wider systemic ramifications: chemically, physically, ecologically and in terms of human health and equity.

Contemporary knowledge should necessarily force us to rethink how we use materials. There are clear challenges entailed in meeting the continuing needs of a growing population wholly dependent on a natural resource base that is greatly diminished, continuing to decline, and also substantially perturbed by the substances we have introduced into it. How we innovate and use materials in a very different tomorrow requires greater regard for their wider ramifications – both positive and negative – if we are to move towards a sustainable and just future.

NOTES

1 United Nations. (2025). *Universal Declaration of Human Rights.* United Nations. [Online.] https://www.un.org/en/about-us/universal-declaration-of-human-rights, accessed 07 January 2025.
2 United Nations. (2025). *The 17 Goals.* United Nations. [Online.] https://sdgs.un.org/goals, accessed 07 January 2025.
3 Maslow, A.H. (1943). A theory of human motivation. *Psychological Review,* 50(4), pp. 370–96.
4 Maslow, A.H. (1954). *Motivation and Personality.* Harpers, New York.
5 Max-Neef, M., Elizalde, A. and Hopenhayn, M. (1989). Human scale development: an option for the future. *Development Dialogue: A Journal of International Development Cooperation,* 1, pp. 7–80.
6 Haberl, H., Fischer-Kowalski, M., Krausmann, F., Martinez-Alier, J. and Winiwarter, V. (2011). A socio-metabolic transition towards sustainability? Challenges for another great transformation. *Sustainable Development,* 19(1), pp. 1–14. https://doi.org/10.1002/sd.410.
7 Lallas, P.L. (2001). The Stockholm convention on persistent organic pollutants. *American Journal of International Law,* 95(3), pp. 692–708. https://doi.org/10.2307/2668517.
8 Kofi Annan, Secretary General of the United Nations on International Day for the Preservation of the Ozone Layer, 16 September 2005.
9 Vitousek, P.M., Aber, J.D., Howarth, R.W., Likens, G.E., Matson, P.A., Schindler, D.W., Schlesinger, W.H. and Tilman, D.G. (1997). Human alteration of the global nitrogen cycle: sources and consequences. *Ecological Applications,* 7, pp. 737–750. https://doi.org/10.1890/1051-0761(1997)007[0737:HAOTGN]2.0.CO;2.

10 Chislock, M.F., Doster, E., Zitomer, R.A. and Wilson, A.E. (2013). Eutrophication: causes, consequences, and controls in aquatic ecosystems. *Nature Education Knowledge*, 4(4), p. 10.

11 A comprehensive review of the microbiome in different groups of organisms is provided by: Appanna, V.D. (2023). *Microbiomes and their Functions: Why Organisms Need Microbes*. Routledge, London.

12 Carabotti, M., Scirocco, A., Maselli, M.A. and Severia, C. (2015). The gut-brain axis: interactions between enteric microbiota, central and enteric nervous systems. *Annals of Gastroenterology*, 28(2), pp. 203–209.

13 Cleveland Clinic. (undated). *Gut-Brain Connection*. Cleveland Clinic. [Online.] https://my.clevelandclinic.org/health/treatments/16358-gut-brain-connection, accessed 19 April 2023.

14 Smitka, K., Papezova, H., Vondra, K., Hill, M., Hainer, V. and Nedvidkova, J. (2013). The role of "mixed" orexigenic and anorexigenic signals and autoantibodies reacting with appetite-regulating neuropeptides and peptides of the adipose tissue-gut-brain axis: relevance to food intake and nutritional status in patients with anorexia nervosa and bulimia nervosa. *International Journal of Endocrinology*, p. 483145. https://doi.org/10.1155/2013/483145.

15 Martinucci, I., Blandizzi, C., de Bortoli, N., Bellini, M., Antonioli, L., Tuccori, M., Fornai, M., Marchi, S. and Colucci, R. (2015). Genetics and pharmacogenetics of aminergic transmitter pathways in functional gastrointestinal disorders. *Pharmacogenomics*, 16(5), pp. 523–539. https://doi.org/10.2217/pgs.15.12.

16 Mace, G.M., Norris, K. and Fitter, A.H. (2012). Biodiversity and ecosystem services: a multilayered relationship. *Trends in Ecology & Evolution*, 27(1), pp. 19–26. https://doi.org/10.1016/j.tree.2011.08.006.

17 George, T.S. (2001). *Minamata: Pollution and the Struggle for Democracy in Postwar Japan*. Harvard University Asia Center, Cambridge, MA.

18 Rutherford, M. (2004). *Institutions in Economics: The Old and the New Institutionalism. Historical Perspectives on Modern Economics*. Cambridge University Press, Cambridge, UK.

19 United Nations Conferences. (1972). *UN Conference on the Human Environment, 5–16 June 1972, Stockholm*. United Nations Conferences. [Online.] https://www.un.org/en/conferences/environment/stockholm1972, accessed 14 March 2023.

20 Meadows, D.H., Meadows, D., Randers, J. and Behrens, W.W. III. (1972) *The Limits to Growth*. Potomac Associates – Universe Books, New York.

21 IUCN–UNEP–WWF. (1980). *World Conservation Strategy: Living Resource Conservation for Sustainable Development*. International Union for Conservation of Nature and Natural Resources (IUCN), Gland. https://portals.iucn.org/library/efiles/documents/WCS-004.pdf, accessed 28 September 20024.

22 WCED. (1987). *Our Common Future*. World Commission on Environment and Development (WCED), Oxford University Press, Oxford.

23 World Health Organization. (2021). *Drowning*. World Health Organization, 27 April 2021. [Online.] https://www.who.int/news-room/fact-sheets/detail/drowning, accessed 11 March 2023.

The Good, the Bad and the Optimal

3

Best judgements of safety matter hugely for environmental sustainability and human health. However, as already observed, a very narrow focus on hazard relating to the intrinsic properties of substances, divorced from the contexts of both environmental behaviour and societal life cycles, can lead to naive judgments about what are 'good' and 'bad' materials. This false dichotomy breaks down rapidly in the light of the use of substances in the real world, taking account of both exposure and hence risk within the broader context of the whole societal life cycles of the products into which they are integrated. This chapter explores aspects of materials often deemed either 'good' or 'bad', extending into different ways of understanding how materials can be considered in terms of their potential suitability for optimally meeting needs in different applications.

3.1 'GOOD' MATERIALS

Naturally occurring and naturally sourced materials are often held up as exemplars of sustainability, with sweeping and misleading but still prevalent assumptions about their safety. Here, we will look at some aspects of naturally occurring and other materials commonly considered safe. This is not to suggest that these often-assumed 'good' or 'benign' materials are automatically a 'bad' thing, but rather to expose some of the unintended negative consequences that can stem from oversimplistic judgments.

In Chapter 2, we considered two of the widespread gases that we breathe every hour of every day. Molecular nitrogen is largely inert and comprises 79% of the air, with oxygen comprising approximately 20% of the lower atmosphere and being essential for life. However, beyond critical partial pressures, nitrogen

DOI: 10.1201/9781003637875-3

also acts as a powerful narcotic and oxygen becomes an instantly fatal neurotoxin. Oversimplistic judgements about what is automatically good or bad without wider consideration can not only be naive but also potentially fatal.

Stone is another example of a basic, naturally occurring and largely inert material. It is interesting that sometimes the 'green lobby' is accused of trying to take society back to the Stone Age. This would not be a very helpful thing, and not just because society's needs would not be met efficiently and societal evolution would hence go into reverse. More significantly, it is also because it is uncritical about the automatic safety of stones. Silicosis, a lung disease caused by inhaling tiny particles of crystalline silica dust, is emerging as a major concern, relating substantially to dust generation from stone and other substances, both natural and synthetic.[1] In fact, fine inhalable silica particles are classified by the International Agency for Research on Cancer (IARC) as a Group 1 carcinogen, meaning that there is sufficient evidence that it causes lung cancer in humans.

Another natural material often deemed benign, and hence 'safe' is wood, as it comes from trees and can in theory be benignly composted or otherwise broken down to be reintegrated into ecological cycles. Ideally, it can be sourced from well-managed forests. However, a consequence of its biodegradability is that products made of wood and wood-based materials in many settings can have a relatively short service life, limiting the satisfaction of human needs over time. Wood products also often require the routine addition of protective agents, particularly biocides, to slow this process and extend functional life. Biocidal application is also potentially hazardous and time-consuming and may result in contaminated waste at the end of life of the product, meaning that benign reintegration into natural systems is no longer possible. Even untreated wood is not automatically safe, as the IARC also classifies wood dust as a Category 1 (proven) carcinogen.[2] This is not to say that wood is a bad thing but to emphasise that an oversimplistic judgement in the absence of real-world use, exposure and risk can lead to sweeping and naive assumptions, including about appropriate precautionary use, that do not help society progress towards sustainability.

3.2 THE RACE TOWARDS 'NATURAL' MATERIALS

Whilst substances produced by biological processes tend to be more benignly integrated back into ecosystems, there are often widespread assumptions and assertions, sometimes explicit but mainly implicit in many assessment tools and approaches, that 'natural is best'. However, there are wider contexts to consider as this touching faith that 'natural' automatically implies 'safe' is hardly borne out when explored objectively.

Dangerous toxins are not restricted to the venom of many snake species, poisons in the skin of frogs such as the groups of 'poison dart' frogs, the sting of a stingray, or curare comprising a range of alkaloid arrow poisons extracted from various plants in Central and South America. Plants and fungi produce a wide range of natural toxins. For example, the toxin ricin produced in the seeds of the castor oil plant (*Ricinus communis*) has an estimated lethal oral dose for humans of approximately 1 mg/kg of body weight.[3]

Plants fungi and sessile animals deploy a wide range of advanced 'chemical warfare' ingredients and other chemically vectored means to compensate for their immobility, both to avoid would-be predators and grazers and to attract mates. Female hop-pickers were known to suffer disruptions to their menstrual cycles due to absorption through the skin of oestrogenic substances contained in the hops that they handle. Many plants, fungi and other organisms also contain or release endocrine-disrupting or otherwise biologically active substances. These are a subset of a far wider array of organisms that produce or contain cocktails of biologically active substances, and many are exploited for their medicinal, toxic and other properties.

3.2.1 Bio-based Materials, Feedstock and Fuel

There is growing interest in biologically based materials. Some derive directly from biological sources, such as cotton and other fibres. Others are processed from plants or other biological materials. The decarbonisation agenda is driving increased interest in a range of plants to meet the growing demand for biologically based chemical feedstocks and also biofuels (generally bioalcohol or biodiesel).

Principal crops serving all of these needs include linseed, canola (oilseed rape), palm oil, soy, sugar cane, wheat and corn. Farmed sources of biological materials are typically classified as: Generation 1 (food-based crops), Generation 2 (non-food crops, which can also include waste from food and forestry crops) and Generation 3 (algal-derived). A primary problem is that it is the food-based (Generation 1) sources of biofuel, feedstock and other materials that have the highest energy density, so there is less demand for substitution with non-food and algal-derived materials. Production of these biologically based materials is not without problems, albeit that these collateral issues are often overlooked by simplistic assumptions.

3.2.2 Bio-based Sourcing

An often-overlooked facet of biologically sourced materials is that they generally do not drop benignly and in profusion from trees with no adverse ramifications. Most, and likely all at the industrial scale, derive from various forms of agriculture. Modern intensive farming is in reality far from an exemplar of benign impacts and bucolic bliss and far from 'natural'. A study published in 2019 by the Intergovernmental Panel on Biodiversity and Ecosystem Services (IPBES) recorded not only that 75% of the terrestrial surface of the Earth had been significantly altered, with farming the largest global contributor to species and habitat loss as well as depletion and degradation of soil contributing to the elimination of in excess of 85% of wetland areas globally compounding water stress in many countries.[4] Agriculture also consumes 70% of the freshwater extracted by people globally.[5] Pressure upon the world's marine environment is no less intense, with the impacts of human activities evident in 66% of oceanic areas.[6] Cumulative appropriation by humanity of the net primary productivity of the world's vegetation stood at between 13% and 25% in 2013, with a projected growth to 27%–29% by 2050.[7] No single species has ever appropriated such a huge proportion of global ecosystem capacity to serve its sole needs.

In fact, when farming, forestry and fisheries are combined with the uses that global society makes of natural resources, humanity consumes the annual productivity of 1.7 'Planet Earths'.[8] Clearly, this is an over-extraction beyond renewable limits, effectively drawing down the capital of the planet's life-support systems.[9] It certainly substantially exceeds 'planetary boundaries' recognised in the Earth ecosystem, beyond which there is an increasing likelihood that abrupt environmental change will occur.[10] So great have humanity's aggregate pressures been upon the structure, integrity and continuing functioning of ecosystems at planetary scale that a seminal scientific paper published in 2000 concluded that we had entered a new epoch, the Anthropocene, in which human impacts have become the dominant factors bringing about ecosystem change, swamping the natural processes that had defined the preceding Holocene.[11] The extraction of such huge amounts of natural productivity, for material production and use as well as food and other purposes, has then to be factored into all life cycle thinking about the supposed benign implications of shifting raw material sourcing from fossil resources towards biologically based alternatives, including far from benign and 'natural' substances produced by intensive farming and forestry.

Cotton fibres are a 'natural' product of the Earth, produced by a variety of species of the genus *Gossypium* (shrubs of the mallow family Malvaceae) naturally occurring across tropical and subtropical regions from the Americas through to Africa, India and Australia. Cotton plants develop protective capsules known as 'bolls' within which seeds mature, and the soft and fluffy fibres surrounding the seeds enable aerial dispersal in exactly the same manner as the 'fluff' produced to promote wind distribution of the seeds of many other plant species such as willow and kapok trees, dandelions and many more. Industrial-scale production of cotton though has significant known sustainability implications. Cotton has been domesticated and produced for human use for millennia, in both the Old World and the New World; the remnants of cotton fabric found in both Mexico[12] and the Indus Valley[13] have been dated to around 5,000 years ago. So how benign is this ostensibly 'natural' material? We have already established that agriculture is the greatest contributor to global losses of biodiversity and habitat as well as soil degradation and water use. Cotton is produced from 2.5% of the world's arable land but has also been found to account for 24% and 11% of the global sales of insecticides and pesticides, respectively.[14] Cotton is also both a water-intensive crop and one growing best in drier countries where water resources are already under substantial pressure, with about 73% of the global cotton harvest coming from irrigated land. The diversion or redirection of water for the growth of cotton has proven environmentally and socially disastrous in some parts of the world. Most strikingly, expansion of the cotton industry in the basins of the Syr Darya and Amu Darya rivers has proven disastrous for the region, leading to massive damage to the Aral Sea, its ecosystem, economy and the health and livelihoods of adjacent communities (see Box 3.1). Cotton also, let us recall, has a chequered social legacy in terms both of its connections with the slave trade when formerly the backbone of the southern American economy, as well as the defiance of Mohandas Gandhi in promoting the spinning of cotton at home across India in non-violent defiance of repressive trade restrictions by British colonial rulers.

BOX 3.1:   COTTON INTENSIFICATION AND THE DECLINE OF THE ARAL SEA

Intensification of cotton farming in the 1960s under the Soviet era has caused massive harm to the ecosystem, the economy and the health and livelihoods of dependent and adjacent communities of the Aral Sea.

In the 1950s, the Aral Sea constituted the world's fourth-largest inland sea supporting a rich fishery and cannery industry. However, state-sponsored intensification of cotton production in drier lands to the north included diversion of the Syr Darya and Amu Darya rivers, major tributaries of the Aral Sea, through massive damming and water transfer schemes. These schemes starved inflows to the Aral Sea that, by 1980, had declined to 10% of its former level, stranding Aralsk City harbour. By 1989, the Aral Sea split into two far smaller hyper-saline seas. By 2003, Aralsk City was separated from the receding water by 64 km (40 miles) of unproductive salt pan.

Drinking water became scarce and extremely polluted by fertilisers and pesticides resulting from cotton farming. Dust storms whipped from the former seabed, now a vast desert, are laced with salt and pesticide residues contributing to high levels of child mortality and respiratory diseases.

Though measures to restore at least a small part of the north basin of the Aral Sea have since been marginally successful, the Sea is a shadow of its former glory days and will likely never recover. The Aral Sea stands as a testimony and warning to the thirst for 'natural' cotton production, a trend sadly also reflected in some other regions of the world promoting the intensive production of cotton.

3.2.3 The Complexities of Assessing the Footprint of Natural Substances

That a substance is biologically sourced is no guarantor that it is benign to nature, nor that its production and impacts are simple. An example is stearic acid, a common and naturally occurring organic acid that is widely used as a raw material for many industrial, food and other purposes, often in the form of a metal-stearate salt. Colourless, odourless and benignly reintegrated into natural cycles, stearic acid may appear to be the ideal 'green' raw material replacing dependence on fossil carbon sources with all their associated problems. Yet, it is necessary to dig a little deeper to explore the full sustainability credentials of its extraction and use.

Stearate can be sourced from both animals and plants. The predominant plant-derived source of stearate today is palm oil. Palm oil plantations have justifiably attracted a great deal of campaigning NGO and media attention over recent years.[15] In addition to the wide range of sustainability pressures already discussed in relation to the scale of intensive farming, palm oil production specifically faces additional environmental and social challenges relating to the drainage and conversion of tropical peatlands for industrial-scale palm oil production. Drainage of these peatlands releases a massive quantity of sequestered carbon as well as changing hydrology at landscape scale, destroying natural habitats with their complement of unique biodiversity, and also displacing formerly forest-dependent tribal communities and lifestyles. Notwithstanding these compounding pressures, including significant climate change implications, the European Union's Renewable Fuels Obligation is placing increasing demands on palm oil plantations to meet a mandatory 10% bioethanol content. Beyond the laudable intent to reduce dependency on fossil carbon in road fuel, a major perverse outcome is the liberation of a cumulatively far greater amount of carbon into the atmosphere relative to fossil fuel use due to wetland drainage and the processing and transport of palm oil derivatives. A Roundtable on Sustainable Palm Oil (RSPO) has been established to certify the supply of palm oil.[16] This, at least in theory, could promote a less impactful raw material supply chain. However, certified volumes are, at the time of writing, small in comparison to demand. Furthermore, the scope and rigours of the certification of this Roundtable are contested.[17]

Alternatively, there is potential to derive stearate from some plant-based waste streams (Generation 2 biomaterials as described earlier in this chapter). If genuinely waste-derived, this has the twin benefits of making beneficial use of material that would otherwise need a disposal route, as well as eliminating demand for palm oil. The use of secondary forestry wastes for creation of a range of bio-attributed polymers is an approach taken by INEOS. (The term 'bio-attributed' indicates that the use of bio-based feedstock is ascribed using the mass balance methodology, rather than that the derived materials are wholly biologically based, for example, as could be verified using radiocarbon methods.) One such example is the bio-attributed polyvinyl chloride (PVC) polymer BIOVYN™ launched in 2019 by Inovyn (part of the INEOS group), made from 100% renewable organic feedstock derived from waste from primary resource extraction that might otherwise require disposal and that does not complete with the human food chain. BIOVYN is claimed to make a major step forward towards carbon neutrality by removing dependency on fossil sources, offering customers a reduction in carbon footprint of over 90% compared with traditional petrochemically derived virgin PVC.[18] This positioned INEOS Inovyn as the world's first commercial producer of bio-attributed PVC driven by commitments to sustainable development. Certification of the biological origins of BIOVYN is independently assessed by the Roundtable on Sustainable Biomaterials (RSB),[19] a global multi-stakeholder membership organisation established to drive sustainable transition to a bio-based and circular economy. INEOS also announced a range of bio-attributed olefins and polyolefins in 2019, equally based on renewable bio-based raw materials that do not compete with food production and that are independently audited by the RSB.[20]

Other companies also offer biologically based and bio-attributed plastics to the market. Just one example is Trinseo, which markets bio-based solutions using second-generation waste such as used kitchen oil and residue from the pulp industry without competing with the food chain, offering complete alternatives to fossil-based plastics or solutions blending biological and fossil sources.[21] Markets for bio-based and bio-attributed plastics are already substantial and still growing. World bio-based plastics production in 2022 totalled 1.9 million tonnes, of which 0.4 million tonnes was produced in the EU27+3.[22] It should not be assumed though that polymers of 'bio-based' origin are biodegradable; many are not due to their chemical structure. In fact, many recyclers refused to accept compostable bioplastics.

Similar considerations apply to other emergent markets for organic acids including, for example, from soybean production. A Round Table on Responsible Soy (RTRS) has been established to enable soy and corn producers of all sizes, serving a range of purposes including for human consumption, animal feed and biofuels, to attain certification as a management tool via a globally recognised sustainable development strategy.[23] A published RTRS Standard for Responsible Soy Production comprises five principles – legal compliance and good business practices, responsible labour conditions, responsible community relations, environmental responsibility and good agricultural practices, aiming to ensure zero deforestation and zero conversion for soy production – linked to 108 mandatory and progressive compliance indicators subject to independent certification.[24] A linked Chain of Custody Standard, also subject to independent certification, describes the mandatory requirements for the different traceability systems an organisation can implement to keep control of RTRS-certified soybean or soy by-product inventories applied across the entire supply chain.[25] The environmental NGO WWF (World Wide Fund for Nature) joined the RTRS in 2010 with the aim of influencing greater sustainability.[26]

Commercial-scale stearate supply can also derive from animal fat. Two routes again apply. One is as a primary product of animal production, for which the full range of agricultural pressures, addressed previously, have to be integrated into sustainability assessment of chemical supply chains. The other route is the beneficial reuse of waste from other primary uses of cattle or other livestock production, such as for food or leather. In practice, this distinction can be hard to make as all animal products contribute to the overall economics of the farming value chain.

One interesting emergent pressure is demand for Vegan chemicals, feedstocks and energy. In theory, this redirects pressure from impacts on animals. In practice, it may instead steer manufacturers away from using animal waste, with its sustainability co-benefits, instead seeking plant-based alternatives for which certified products are in limited supply. It may thereby inadvertently increase demand for virgin palm oil with its host of biodiversity, hydrological, climate forcing and social implications, all of which have massively net negative impacts affecting animals as well as people at all scales from the local to the global.

3.2.4 Natural Production of Problematic Substances

Another of the common assumptions implicitly underpinning the misplaced 'good' versus 'bad' material dichotomy is that all of the most potentially problematic categories of 'bad stuff' are synthetic. As one example, there is an often-stated yet flawed assumption that chlorinated organic and other halogenated organic substances are all human-made. The reality is that various microbial life forms produce halogenated organic molecules for anti-fouling (particularly in sponges) as well as wider biocidal and other purposes. Microorganisms are significant producers of halogenated natural products.[27]

Bacteria and fungi, in particular, play significant roles in the production of these compounds that serve a range of functions such as reducing palatability as a defence mechanism or to kill grazers or predators or otherwise deter attack. Penicillin, for example, is a halogenated antibiotic produced by the *Penicillium* fungi. Bacteria of the genus *Bacillus* are also known to produce halogenated compounds including antibiotics and biocides. Many microbially produced halogenated compounds are persistent due to their chemical structure, resisting degradation by natural processes. Marine algae including both macroalgae and microalgae, for example, have been found to produce a wide range of halogenated metabolites, including various acyclic and polycyclic chlorinated molecules that are receiving attention for medical applications due to their biologically active properties.[28]

Over 5,000 natural organohalogen compounds have been identified, along with at least five distinct halogenating enzyme systems.[29] The bulk of these natural organohalogen compounds in terrestrial environments is integrated with humic polymers, whilst naturally occurring chlorinated and brominated bipyrroles as well as methoxypolybrominated phenyl ethers are naturally produced and can be widespread in marine environments and are biomagnified in sea mammals. Evidence suggests that the cycling of halogens such as chlorine and bromine in soils is largely driven by microbial processes, particularly by bacteria, fungi and archaea, with halogenation and dehalogenation of organic matter constituting part of normal and natural cycling within soils.[30] Natural sources of organochlorine substances primarily dominate the global budget of chloromethane and chloroform.

Methylation of metals, such as the production of various species of methylmercury, is also a natural process albeit amplified massively by anthropogenic pollution with mercury-rich wastes. Add to this the diversity of neurotoxic, hepatotoxic and haemotoxic products released by blooms of cyanobacteria particularly in nutrient-enriched freshwater lakes, and the generation of other toxins in marine dinoflagellate blooms leading to cases of paralytic shellfish poisoning when filter-feeding organisms such as mussels are consumed by people after an algal bloom. Suddenly, the concepts of 'natural' and 'safe' seem far from uniformly comfortable bedfellows.

3.2.5 Take-Home Messages About Bio-based Materials

The purpose of this discussion of 'natural' materials – and let us not forget that intensive farming is far from a natural activity into which substantial energy and chemical inputs are often overlooked – is not to suggest that substances derived from 'natural' products are 'bad' or indeed automatically 'good'. Many have substantial contributions to make to sustainable development in the right contexts, as exemplified, for example, by bio-attributed polymers derived from agricultural and forestry wastes, or indeed paper that is recycled at a high rate in many regions. The key take-home message here is that everything must be evaluated in context.

Simplistic judgments about what is 'good' or 'safe', or 'bad' or 'detrimental', can cause problems if applied in too incautious a way. Biologically derived substances may be used beneficially but are certainly in no way automatically a panacea compared to synthetic counterparts. There is a need to think deeper, and that is what we will turn to in the following chapters.

3.3 'BAD' MATERIALS

By contrast, some persistent substances may automatically be branded as 'bad'. In some instances, such as the wider environmental release of toxic and persistent nerve agents, pesticides and damaging metals such as mercury or cadmium, that may be warranted. Some other substances have a higher degree of hazard associated with them, for example uranium or cadmium, the toxicity of which is so significant and the likelihood of exposure through incautious use so high that they have to be contained tightly, as for example in a nuclear reactor (though history from Three Mile Island to Chernobyl and Fukushima as well as vulnerability as terrorist or military targets raises concerns even there). But these materials are not suitable for many wider uses from which human or environmental exposure is an inevitability.

Other substances are undoubtedly hazardous when released into the human body or the wider environment. Yet contained or careful management, or the location in which these substances occur, can radically change the balance of harm and benefit. One such example is that every person produces hydrochloric acid in their stomach at a pH of between 1.5 and 3.5. Hydrochloric acid is distinctly hazardous in any other body organ or environmental location. It is also the material – gaseous muriatic acid – for which the Alkali Act was initially created in the UK in 1863. However, hydrochloric acid is produced locally in the stomach where it is needed, serves a vital metabolic function and is then safely resorbed without exposure elsewhere within the body let alone in the wider environment. Another example is ozone, a known carcinogen (IARC Category 1) and contributor to formation of smog at low altitudes, yet ozone has many applications such as sterilisation of water under controlled conditions and is also a vital constituent of the 'ozone layer' of the stratosphere absorbing most of the damaging ultraviolet radiation arriving from the sun. Context, including exposure, is everything.

As already discussed, though wood is lazily described as a 'good' material', wood dust is a proven carcinogen classified by the IARC as Category 1. Stringent handling of wood dust in manufacturing sites can render the substance safe and available for beneficial reuse, though environmental exposure is far less rigorously controlled than for other chemical substances. Simplistically, if plants, fungi and other biological resources are automatically assumed to be 'good', what about the toxins and other biologically active agents extracted or deriving from them? Also, the prevalence of silicosis amongst miners and other workers routinely exposed to stone dust. Where is the clear line between these very different prejudgments and potential impacts?

Synthetic substances receive far deeper scrutiny than this. The manufacture of many complex chemicals depends on constituents that may themselves be manufactured, including both primary materials and additive substances contributing beneficial functionality. The presumptions about and regulatory approach to these synthetic moieties are different from materials that may be hazardous yet derive from biological or mineral resources or that may have been used throughout history. This is not to say that hazard should be disregarded; far from it in fact. But, as one example, vinyl chloride monomer (VCM) is an IARC Category 1 (proven human carcinogen) polymerised into PVC and so is analogous with the handling of wood dust (also IARC Category 1), but PVC manufacturing facilities are now tightly controlled avoiding human environmental exposure. In what way does that differ from the optimal handling of wood dust, which is in the main far from tightly controlled? Stringent controls in manufacturing processes today ensure containment of chemical precursors, eliminating exposure and meaning that intrinsic hazard does not translate to actual risk due to prudent and legally mandated mitigations.

Where substances are inert and durable, as for example the use of 'persistent' plastics that are chemically inert and resistant to decay when used as window profiles, pipes, wiring insulation, ICT infrastructure, flooring or other long-life applications, they can deliver a long service life and hence a fulfil a high level of human needs per unit of resource, also requiring no or minimal maintenance inputs throughout life. Many plastics and metals are then recoverable for cyclic reuse or remanufacture once products have reached the end of their service lives.

Other known hazardous substances are widely used in preservative or biocidal applications with the primary purpose of arresting biological processes leading to degradation so, by definition, might be assumed to be 'bad' materials. And yet they serve beneficial purposes in prolonging the shelf life and safety of food or prolonging the beneficial service life of finished products, hence contributing to their efficiency in meeting needs. Akin to the use of medicines and pesticides, for which therapeutic and (known) detrimental properties are balanced in making decisions about application, 'bad' properties can have 'good' contributions to meeting human needs if used and handled wisely.

These contrasting examples of wood and plastics, whilst not a universal reality for all applications of these materials, illustrate the importance of thinking at the scale of whole product life cycles, taking account of factors such as durability and maintenance inputs throughout service life as well as their potential for recovery and recyclability. This wider conceptual framework challenges simplistic and false dichotomies automatically casting judgments about 'good' versus 'bad', or of 'better' or 'worse', materials when their use is set within the wider framing of sustainability, contextualised across whole product life cycles.

Context, once again, is everything.

3.4 SCIENCE, SIMPLICITY AND NAIVETY

The deeper we dig beneath the surface of the culturally constructed dichotomy of what is assumed to be uniformly 'good' or 'bad', the less helpful such judgements seem to be in propelling society in a more sustainable direction. The key lesson emerging from this observation of narrow judgments about what constitutes 'good' versus 'bad' materials is that context is everything, with no clear line demarking either presumption. Furthermore, the most significant differentiator lies not necessarily in the inherent properties of the substances themselves, although these must be taken seriously into account. Rather, for both purported 'good' and 'bad' materials alike, it is a matter of how potential hazard translates through exposure into actual risk in the context of 'real-world' use.

A key sustainability question is how the use of substances – all substances evaluated objectively across product life cycles – can help us fulfil our needs in the safest and most efficient manner. Context and balance, with mitigation as necessary, are everything.

NOTES

1   Hoy, R.F., Jeebhay, M.F., Cavalin, C., Chen, W., Cohen, R.A., Fireman, E., Go, L.T.H., León-Jiménez, A., Menéndez-Navarro, A., Ribeiro, M. and Rosental, P.-A. (2022). Current global perspectives on silicosis—convergence of old and newly emergent hazards. *Respirology*, 27(6), pp.387–398. https://doi.org/10.1111/resp.14242.

2   IARC. (1995). *Wood Dust and Formaldehyde: IARC Monographs on the Evaluation of Carcinogenic Risks to Humans*, Volume 62. [Online.] https://publications.iarc.fr/Book-And-Report-Series/Iarc-Monographs-On-The-Identification-Of-Carcinogenic-Hazards-To-Humans/Wood-Dust-And-Formaldehyde-1995, accessed 15 November 2024.

3   European Food Safety Authority. (2008). *Ricin (from Ricinus communis) as undesirable substances in animal feed. Scientific opinion of the Panel on Contaminants in the Food Chain*, *EFSA Journal*, 6(9), p.726. https://doi.org/10.2903/j.efsa.2008.726.

4   IPBES. (2019). *Global Assessment Report on Biodiversity and Ecosystem Services*. Intergovernmental Panel on Biodiversity and Ecosystem Services (IPBES). [Online.] https://ipbes.net/global-assessment, accessed 25 January 2022.

5   Altobelli, F., Cimino, O., Natali, F., Orlandini, S., Gitz, V., Meybeck, A. and Dalla Marta, A. (2018). Irrigated farming systems: using the water footprint as an indicator of environmental, social and economic sustainability. *Journal of Agricultural Science*, 156(5), pp. 711–722. https://doi.org/10.1017/S002185961800062X.

6   IPBES. (2019). *Global Assessment Report on Biodiversity and Ecosystem Services*. Intergovernmental Panel on Biodiversity and Ecosystem Services (IPBES). [Online.] https://ipbes.net/global-assessment, accessed 25 January 2022.

7   Krausmann, F., Erb, K.-H., Gingrich, S. and Searchinger, T.D. (2013). Global human appropriation of net primary production doubled in the 20th century. *Proceedings of the National Academy of Sciences of the United States of America*, 110(25), pp. 10324–10329. https://doi.org/10.1073/pnas.1211349110.

8   Global Footprint Network. (2020). *Living Planet Index 2016*. Global Footprint Network. [Online.] http://www.footprintnetwork.org/en/index.php/GFN/, accessed 25 January 2022.

9   Raworth, K. (2018). *Doughnut Economics: Seven Ways to Think Like a 21st-Century Economist*. Random House Business, London.

10 Rockström, J., Steffen, W., Noone, K., Persson, Å., Chapin, F.S., Lambin, E., Lenton, T.M. Scheffer, M., Folke, C., Schellnhuber, H., Nykvist, B., De Wit, C.A., Hughes, T., van der Leeuw, S., Rodhe, H., Sörlin, S., Snyder, P.K., Costanza, R., Svedin, U., Falkenmark, M., Karlberg, L., Corell, R.W., Fabry, V.J., Hansen, J., Walker, B., Liverman, D., Richardson, K., Crutzen, P. and Foley, J. (2009). Planetary boundaries: exploring the safe operating space for humanity. *Ecology and Society*, 14(2), p. 32.

11 Crutzen, P.J. and Stoermer, E.F. (2000). The 'Anthropocene'. *Global Change Newsletter*, 41, pp. 17–18.

12 Huckell, L.W. (1993). Plant Remains from the Pinaleño Cotton Cache, Arizona. *Kiva, the Journal of Southwest Anthropology and History*, 59(2), pp. 147–203.

13 Moulherat, C., Tengberg, M., Haquet, J.F. and Mille, B. (2002). First evidence of cotton at Neolithic Mehrgarh, Pakistan: analysis of mineralized fibres from a copper bead. *Journal of Archaeological Science*, 29(12), pp. 1393–1401.

14 Worldwide Fund for Nature. (2000). *The Impact of Cotton on Fresh Water Resources and Ecosystems*. Worldwide Fund for Nature. [Online.] http://wwf.panda. org/?3686/The-impact-of-cotton-on-fresh-water-resources-and-ecosystems, accessed 26 May 2020.

15 WWF-UK. (n.d.). *8 Things to Know about Palm Oil*. WWF-UK. [Online.] https://www.wwf.org.uk/updates/8-things-know-about-palm-oil, accessed 03 October 2024.

16 RSPO. (204). *A Global Partnership to Make Palm Oil Sustainable*. Roundtable on Sustainable Palm Oil (RSPO). [Online.] https://rspo.org/, accessed 18 October 2024.

17 Browne, P. (2009). Defining 'Sustainable' Palm Oil Production. *The New York Times Blogs*, 6 November 2009. [Online.] https://archive.nytimes.com/ green.blogs.nytimes.com/tag/roundtable-on-sustainable-palm-oil/, accessed 03 October 2024.

18 INEOS Inovyn. (2023). *BIOVYN – INOVYN Bio-Attributed PVC: A Big Step Forward towards Carbon Neutrality*. INEOS Inovyn. [Online.] https://biovyn. co.uk/, accessed 14 May 2023.

19 RSB. (2023). *What is the RSB?* Roundtable on Sustainable Biomaterials (RSB). [Online.] https://rsb.org/, accessed 13 May 2023.

20 INEOS. (2019). *INEOS Olefins & Polymers Europe Announces Range of Bio-Attributed Olefins and Polyolefins*. INEOD Press Release, 17 October 2029. [Online.] https://www.ineos.com/news/shared-news/ineos-olefins--polymers-e urope-announces-range-of-bio-attributed-olefins-and-polyolefins/, accessed 26 September 20024.

21 Trinseo. (2024). *Bio-Based, Bio-Attributed and Bio-Degradable Plastics*. Trinseo. [Online.] https://www.trinseo.com/Solutions/Bioplastics-Biodegradable-Plastics, accessed 28 October 2024.

22 Plastics Europe. (2024). *Bio-Based and Biodegradable Plastics*. Plastics Europe. [Online.] https://plasticseurope.org/sustainability/climate/circular-feedstocks/ bio-based-and-biodegradable-plastics/, accessed 28 October 2024.

23 RTRS. (2024). *What are the Benefits of RTRS Certification?* Round Table on Responsible Soy (RTRS). [Online.] https://responsiblesoy.org/certificacion? lang=en, accessed 23 September 2024.

24 RTRS. (2021). *RTRS Standard for Responsible Soy Production V4.0*. Round Table on Responsible Soy (RTRS). [Online.] https://responsiblesoy.org/documentos/rtrs-standard-for-responsible-soy-production-v4-0?lang=en, accessed 23 September 2024.
25 RTRS. (2021). *RTRS Chain of Custody Standard V2.3*. Round Table on Responsible Soy (RTRS). [Online.] https://responsiblesoy.org/documentos/rtrs-chain-of-custody-standard-v2–3?lang=en, accessed 23 September 2024.
26 WWF. (2010). *WWF, The Round Table on Responsible Soy and Genetically Modified Soy*. WWF. [Online.] https://www.wwfca.org/en/?156602/Involvement-in-the-RTRS-GM-Soy-Industry, accessed 03 October 2024.
27 Murphy, C.D. (2010). Halogenated organic compounds – carbon-halogen bond formation. In: Timmis, K.N. (ed), *Handbook of Hydrocarbon and Lipid Microbiology*. Springer, Berlin, Heidelberg. https://doi.org/10.1007/978-3-540-7 7587-4_25.
28 Cabrita, M.T., Vale, C. and Rauter, A.P. (2010). Halogenated compounds from marine algae. *Marine Drugs*, 8(8), pp. 2301–2317. https://doi.org/10.3390/md8082301.
29 Field, J.A. (2016). Natural production of organohalide compounds in the environment. In: Adrian, L., Löffler, F. (eds) *Organohalide-Respiring Bacteria*. Springer, Berlin, Heidelberg. https://doi.org/10.1007/978-3-662-49875-0_2.
30 Weigold, P., El-Hadidi, M., Ruecker, A., Huson, D.H., Scholten, T., Jochmann, M., Kappler, A. and Behrens, S. (2016). A metagenomic-based survey of microbial (de)halogenation potential in a German forest soil. *Scientific Reports*, 6, p. 28958. https://doi.org/10.1038/srep28958.

Material Risks Over Whole Societal Product Life Cycles

4

From simplistic judgments about 'good' versus 'bad', or the evaluation of safety on the basis solely of intrinsic chemistry in the absence of real-world risks, we've seen that misinformed judgments can often occur about the relative safety and efficiency of material use. This includes both risks averted or mitigated, for example when substances are fully contained in intermediate processes or cyclic reuse. Examples given include hydrochloric acid, an undoubtedly hazardous material to human health and the environment, yet beneficially produced and contained and then wholly consumed or resorbed as an essential agent in digestion and decontamination processes in the stomachs of humans and many mammals. The same principle of risk management or mitigation applies, or should do, to the industrial use of substances contained in controlled situations that then do not manifest in finished products, and so averting risks to people and nature. Examples given of beneficial, controlled uses included ozone (an IARC Category 1 proven carcinogen substance) in the treatment of swimming pools and wastewater, as well as in many industrial processes producing materials from which products supporting a diversity of human needs. Extending downstream in material use life cycles, substances such as copper, a known aquatic toxin, are in very wide use in plumbing, wiring and other applications and yet are efficiently recovered and recycled limiting accumulation in natural systems.

DOI: 10.1201/9781003637875-4

45

4.1 WHY A LIFE CYCLE PERSPECTIVE MATTERS

A whole societal product life cycle context expands consideration to many commonly overlooked inputs and outputs entailed in maintaining longevity and function, be that to synthetic or ostensibly 'natural' or 'safe' materials. We have already looked at examples of wood and other biologically derived materials that not only tend to deliver a reduced level of human need through lower durability and hence shorter serviceable life but also require routine inputs of substances to slow their degradation that, in turn, may inhibit or entirely prevent recycling and value recovery.

Taking a systemic perspective, it is also necessary to account for various additives that may be introduced into primary materials during whole societal product life cycles. Additives may be included as constituents of materials such as in polymers, in the pretreatment of wood with preservatives or in the formation of alloys with other metals. Other additives might be introduced at a compounding or a converting stage as complex materials are synthesised, added for purposes such as extending product functionality, or as processing aids facilitating processes at the manufacturing stage. And then there are the often-overlooked additives applied throughout the use phase of products as, for example, considered previously in the context of preservatives added to wooden products, as well as water and other chemical and labour inputs for maintenance over what might be extended product life. In the examples that follow, the sustainability footprints of these additives introduced during the product use phase can sometimes dwarf the initial sustainability footprint of the primary material. This emphasises the importance of a fully systemic perspective when evaluating the sustainability of material use, as indeed the myopia of the naive and unhelpful judgments when too narrow a view is taken.

A further aspect of a systemic overview of the use of materials over whole product life cycles is that some additives might be introduced at early synthesis or compounding stages, apparently prejudicing the sustainability credentials of the primary material, whilst for other materials the sustainability burden of additives might be introduced later during product use or other life cycle phases and may be highly significant. Some facilitate while others obstruct recyclability and hence progress towards materially, energetically and socially beneficial circular resource use. A whole life cycle perspective is vital for objective evaluation of the full sustainability footprint of the use of materials, including both their positive contributions to meeting needs and risks requiring management.

4.2 LIFE CYCLE ASSESSMENT

Having determined the naivety of considering the merits and perils of materials based narrowly on intrinsic properties outside of the context of whole product life cycles, and hence actual risks, let us then turn to some considerations germane to their sustainable use across these life cycles.

The international standard ISO 14040 simplifies life cycle thinking and is widely applied globally in industrial applications. The simplified life cycle stages consistent with ISO 14040 therefore form a logical basis for looking at the different implications of material use in products across the societal life cycle. In essence, this simplification is into six life cycle stages running from: (1) sourcing of raw materials; (2) synthesis of primary and additive materials; (3) packaging and distribution of those material products; (4) compounding of different materials and the subsequent conversion of this mix or compound into finished products; (5) the use phase of finished products including associated inputs and outputs; and (6) the post-use phase be that cyclic recovery, reuse or recycling or some form of disposal (see Figure 4.1). The following subsections address each of these life cycle stages in turn, with examples of material behaviours germane to the sustainability implications of their use.

4.2.1 The Raw Material Extraction Phase

The raw materials used in chemical synthesis derive from a range of geographical regions and sources, significantly including mined substances and also those derived from agriculture and forestry. Recycled materials can also be utilised as inputs to the life cycle.

A range of associated impacts include the potential for chemical pollution through mining activities and collateral releases of substances, emissions of climate-active gases particularly related to energy use, physical impacts on the locations from which raw materials are extracted affecting biodiversity as well as aquifers, and also human impacts associated both with labour and neighbouring communities. To cover all these aspects in detail would be to overburden this chapter. A few generic considerations are therefore addressed here, along with some practical examples.

Mining is obviously a disruptive way of extracting materials, with inevitable consequences for soils and underlying geology as well as associated aquifers and ecology. Initiatives have been taken to minimise these impacts or to mitigate damage where inevitable, which is certainly progress measured

FIGURE 4.1 Simplified societal life cycle from: raw material extraction, chemical synthesis, packaging and distribution, compounding and converting into products, product use, and post-use optimally leading to circular recovery and reuse.

against prior extraction without regard for these considerations. Mining of lithospheric substances as raw materials and for energy generation, such as metals, nutrients and fossil carbon, is inherently problematic as these materials have been incrementally deposited and 'locked away' from the biosphere by physicochemical, depositional and biomineralisation processes over a period of billions of years. Humanity has only been present for approximately 0.004% of the period over which these lithospheric processes have been operating, and acceleration of mineral exploitation since the onset of the European Industrial Revolution comprises just 0.00004% of planetary history. However, the pace of this geologically recent remobilisation of sequestered substances into biospheric cycles has created a host of known problems – metal toxicity, eutrophication, radiation build-up and climate change amongst them – and may contribute additionally to unforeseen problems beyond currently known thresholds. This consideration is germane to the raw material extraction phase, but it also ramifies across all downstream life cycle stages if these mined materials are not recaptured in technical cycles to prevent their diffusion into ecosystems and consequent systematic accumulation leading to both known and as-yet-unknown problems. Within the context of whole life cycle thinking, the avoidance or minimisation of impacts at this stage is obviously desirable. However, in the real world, full cognisance of these impacts needs to be brought to bear in sustainability assessment and, where possible, measures need to be sought to avoid or mitigate adverse impacts.

Ethical issues in addition to environmental consequences are also inherent in raw material extraction through mining. A well-established example here is 3TG substances – tin, tantalum, tungsten and gold – significant sources of which derive from places associated with repressive regimes and damaging extraction processes, for example including the Democratic Republic of Congo where child and indentured labour is common. For this reason, the Responsible Minerals Initiative has developed a Conflict Minerals Reporting Template (CMRT) against which organisations can assess whether their use of these minerals derives from repressive regimes, self-reporting this openly on the CMRT as a transparent declaration about ethical responsibility to their downstream customers.[1]

Mining may also displace important and supportive habitats whilst also disrupting aquifers and displacing associated local rights and livelihoods. Mined extraction of persistent substances that may be hard to control throughout the life cycle, such as heavy and naturally scarce metals of major concern (examples include lead, cadmium and mercury), can also create inherent risks as well as result in the incidental environmental release of co-contaminants. Dependence on fossil carbon-based fuels for raw material extraction processes, as well as releases of other sequestered materials during the extraction process, can also create environmental problems as well as human health risks. These risks can be revealed and considered through responsible and ethical sourcing policies and practices addressing the potential for chemical pollution, habitat, water and other natural resource degradation, as well as ethical consideration for employees as well as affected communities including fair allocation of risks and rewards. Traceability of raw materials is increasingly important for users further down the value chain.

As extensively discussed in Section 3.2, assumptions that sourcing of biologically based raw materials is automatically more sustainable, be that from agriculture, forestry or intensive fishing, are rarely realistic given the massive magnitude of pressures that these activities impose on dwindling ecosystems. Various approaches to sustainable forestry are widely established, for example entailing block-based felling to provide refuges and mitigation for biodiversity as well as replanting for renewability and respect of forest-dwellers and their livelihoods. Chapter 7 covers the purpose and uptake of the Forest Stewardship Council (FSC) brand[2] across value chains associated with forest-derived products. As we have also seen in Chapter 3, a Roundtable on Sustainable Palm Oil has been established to promote and certify the supply of palm oil in compliance with some sustainability considerations, and a Round

Table on Responsible Soy has also been established with similar aims.[3] The fact remains though that naive assumptions persist that 'biologically based' material production is automatically benign, when we know that agriculture is the most significant factor globally for loss of biodiversity and habitats, utilisation of abstracted freshwater as well as erosion or degradation or soil, with industrial farming for feedstock and biofuels also displacing local communities and their food, traditions and livelihood needs.

A particularly striking statistic is that around 2,700 L of fresh water, equivalent to a single person's drinking needs for 900 days, are required to make a single cotton tee-shirt, much of this related to the water intensity of cotton grown as a crop.[4] This compounds the wider pressures imposed by cotton production, as reviewed in the previous chapter. The Better Cotton Initiative seeks to engage whole cotton value chains from producer organisations, suppliers and manufacturers, retailers and brands, not-for-profit organisations and wider associated groups in more sustainable cotton supply and use favourable to smallholders, larger farms and farming communities.[5] But a sustainability footprint – environmental and social and with economic implications – is inevitable and needs to be factored into the evaluation of products derived from or containing cotton as a raw material.

Energy use and its implications for climate-active aerial emissions, particularly of carbon dioxide released from fossil-based energy, are a significant part of the sustainability footprint in the raw material extraction phase. Whilst these emissions may be offset to a certain degree by energy efficiency, a pivot towards renewable sources or mitigation measures, they nevertheless form a significant part of the overall carbon audit of product life cycles. With increasing focus on climate change as a regulatory and business consideration, auditing of climate-active emissions associated with energy and other activities at all points in the value chain will increasingly become the norm. This creates incentives for more energy-efficient or low-carbon approaches favourable for the selection of raw material suppliers who are increasingly focused on net zero goals driven by best intentions, cost reduction and avoidance of carbon taxes, potentially tightening regulations, demand from companies further down value chains that revise environmental and social aspects of their sourcing strategies, consumer demands or expectations and the likelihood of increasingly tightening regulation.

4.2.2 Material Synthesis

Images of factories spewing clouds of gases are commonplace in campaigning literature. Emissions to all environmental media – water, air and soil – as well as waste reduction and elimination and also health and social consequences for workers and neighbours, remain issues to be addressed. Factors such as chemical and energy inputs to the manufacturing stage and potential generation and handling of by-products, waste and emissions need to be carefully controlled and audited. Targets to achieve net neutral impacts on ecosystems and communities, including ethical employment policies, are important.

The material synthesis phase has seen major focus and action, both statutory and elective, even from the days preceding the Alkali Act. Often, for all that it remains the target of media and non-governmental organisation (NGO) attention, the synthesis phase of material life cycles is generally the most rigorously controlled. By contrast, other phases in the total societal life cycle beyond manufacturers' gates have, until relatively recently, been perceived or dismissed as problems to be dealt with by other people and sectors of society.

This culture of focusing only as far as the factory gate is still seen, for example, within the European requirement to issue Safety Data Sheets documenting substances of potential concern and advice to customers on safe storage and use of manufactured chemicals. Of course, the functional contribution of substances in the compounding and converting phases, as well as in the product use phase not to mention material design to facilitate or at least prevent obstacles to recycling, is germane to material formulation well beyond the factory gate. Continued scrutiny of all inputs, outputs and implications – chemical, physical and social – though remains essential for continuous improvement in the material synthesis phase of the life cycle.

4.2.3 Material Packaging and Transport

If the material synthesis phase is one of the more tightly controlled aspects of the life cycle due to historic pressures, the packaging and transport phase as material products are shipped beyond the factory gates is, in my experience working with many major chemical companies, one of the most overlooked. There are different facets to be considered. This includes the packaging type deployed and its sourcing. Also, the ultimate fate of packaging materials, whether recovered and returned, recycled or disposed of in general waste streams.

Packaging materials also have their own supply chains. These are often poorly considered in relation to the scrutiny of the sourcing of other raw material inputs to chemical synthesis. Following its role in safely containing the chemical products, is the material from which the packaging made recoverable, and ideally actually recovered, for either benign disposal or, better still, reuse or recycling? Shipping of materials in industrial bulk containers (IBCs) can be efficient where the IBCs are part of take-back and reuse schemes. Other materials may be transported in bulk in tankers averting the need for separate packaging, offering significant advantages not only by minimising packaging materials and waste generation but also by enhancing transport efficiency per unit of delivered material. Questions germane to packaging materials include whether they contain problematic substances such as scarce metals or other persistent substances that may not be recovered for reuse or benign disposal. Is it possible to create closed loops, either through returnable and reusable packaging or else design such that packaging materials can serve further beneficial uses beyond the delivery of their material contents? As a minimum, measures to ensure that packaging does not enter landfill, incineration or other inefficient disposal processes should be considered.

Then, there is the transport mechanism. By and large, transport systems established around the world are acknowledged as unsustainable, with a high dependence on fossil carbon and physical infrastructure. The road network that we have inherited is a symptom of a global system about which the material manufacturer shipping their product may have little control. This is not though to say that the manufacturer can't have some influence through their transport supplier policies with regard to practises covering worker ethics and selection of the least materially intense and damaging transport mode. Shipping in bulk clearly minimises impact per unit of material product shipped that, for some major customers, may be a preferred method of delivery reducing the footprint for both players in the value chain.

In a world in which increasing attention is paid by regulators and customers to climate-active emissions associated with energy, transport can be a significant contributor to the overall sustainability footprint of the total product life cycle. It therefore needs, as a minimum, to be audited, and then ideally reduced or mitigated as far as feasible.

4.2.4 Compounding and Converting

The manufactured material is then bought by further companies that compound it into more complex materials, with a further step then converting this compounded material into final products. Further chemical and energy inputs can give rise to sustainability concerns at the compounding and converting stages including, for example, inputs of persistent metals or organic substances that may be released as emissions or may become mobile in the compound. Further biologically sourced materials may be introduced at this stage, requiring transparent consideration as noted previously. Careful consideration of formulations has therefore to be brought to bear to ensure that these risks are minimised or averted. Energy considerations, again, can be significant in these compounding and converting stages and need to be factored into the overall sustainability footprint of the product life cycle.

Human impacts – health implications for workers and nuisance as well as potential risks to neighbouring communities – need to be integrated centrally into the management strategies of compounding and converting companies. As with the material synthesis phase, these further manufacturing stages in the life cycle have also generally been under intensive scrutiny by regulators, NGOs and the media, meaning that they tend to be tightly controlled both through regulation and self-interest. There nonetheless remains scope for continuous improvement across all chemical, physical and human dimensions of sustainability.

4.2.5 The Product Use Phase

As observed in previous chapters, historic scrutiny of materials has generally focused on the potential for negative outcomes and particularly consideration of hazards in isolation from real-world contexts. However, materials are not innovated and used to create problems, but to address human needs; it is the benefit that material use confers that is the basis for their selection and innovation. These positive contributions of material use are particularly evident in the use phase of products designed to serve a wide diversity of needs.

Attributes of materials such as longevity of service life have positive sustainability virtues, enabling the products made from them to support needs efficiently per unit of physical resource. This principle is often overlooked in material assessment, both when considering primary materials but also additive substances compounded with them to enhance durability and resilience as well as imbuing compounds with a wide range of functional properties. From the perspective of sustainability assessment, longevity is a particular virtue in terms of conferring the greatest value per unit material resource in the meeting of needs.

Outcomes at the product use phase are often key considerations in formulation during the material synthesis phase, as well as influencing the raw material extraction phase in terms of either substituting potentially hazardous raw material inputs or ensuring that these are fully consumed in the synthesis phase. Product use considerations also inform the combination of materials in the compounding and converting phases to achieve the desired functional attributes.

Material behaviour during the use phase also needs to be considered in terms of the potential for substances to enter and accumulate in the wider environment, including into humans whether by leaching out, through chemical breakdown, or via mechanical degradation such as dust formation. Potential hazard needs to be contextualised with actual exposure, and hence 'real-life' risks. If there is a risk, some form of mitigation, or potentially material substitution, may be necessary for that application.

Beyond consideration of material behaviour during the product use phase, additional inputs and outputs entailed in product maintenance are often overlooked yet may be significant. This is particularly the case for products with a long service life. As outlined previously, sourcing of some natural materials such as timber may appear to be sustainable when viewed purely in the context of stewardship of productive forests, but they may require substantial inputs of preservatives such as paints, varnishes or biocides to maintain function. Other maintenance inputs potentially include substances such as water and cleaning materials as well as energy and labour. These inputs into the product use phase may potentially dominate the overall sustainability footprint in long-life applications. When considering lazily assumed 'good' and 'bad' materials earlier in this book, it was noted that window profiles made from durable materials require no such maintenance inputs in a far extended use phase beyond which the material can be (and largely is the case in Europe) recovered for mechanical recycling with virtually no loss of technical properties. Durable materials can thereby reveal substantial potential contributions to overall sustainability in the product use phase, particularly when recycled for further beneficial use cycles.

In a world of rising human numbers, compounded by increasing per capita resource consumption (ironically a positive legacy of development programmes as a greater proportion of humanity is brought out of poverty), there is an obvious pressing need to satisfy more human needs per unit of physical substance when one considers the seriously declining trends apparent in biodiversity, soil and other natural resources. Durability emerges as a key component in the sustainable development journey, with the longevity of products and the maximisation of their service life constituting a fundamental contributor to resource efficiency. It is in this context that the persistence of substances, often perceived or represented as a negative quality, comes to the fore in terms of the durability of the products from which they are manufactured or within which they are integrated. This is particularly important for long-life applications such as those used in built infrastructure – pipework, wires, window profiles, membranes and flooring are good examples – where biodegradation of less resistant materials can present significant problems. The durability of materials such as various forms of plastic, metal and ceramic offers prolonged service life without the need for frequent maintenance with associated energy and chemical inputs, and more frequent replacement and associated disruption, constituting far more efficient delivery against human needs per unit of physical resource. Of course, it matters that the material in question is inert in use, and ideally that it is recoverable for beneficial reuse, avoiding environmental accumulation which remains a priority further downstream in the life cycle.

By contrast, natural materials have obvious benefits in short-life applications, such as food packaging or writing materials. Paper, for example, can be safely reintegrated into natural cycles through composting and other means assuming, of course, that it is not coated, does not contain additives that inhibit benign reintegration or is contaminated in use. As addressed in the following section, these factors contribute to a significant recycling rate of paper across Europe.

All inputs and outputs need to be recognised in assessing contributions – both positively in terms of meeting needs but also negatively – during the use phase within the overall societal life cycle of products. The water intensity of cotton production was highlighted in consideration of raw material extraction, citing the water consumed to produce a single cotton tee-shirt. Another stark statistic relating to the use phase of this commonplace garment is that simply air-drying a cotton tee shirt rather than using a drier can save one-third of its total product life cycle carbon footprint.[6] The design and use of materials of all types can make positive contributions if the use phase is extended with reduced inputs of all types. This observation also relates to so-called 'smart materials' such as photovoltaic, temperature-responsive, self-healing and chemo-responsive materials that respond to environmental changes, leading to more efficient resource usage and energy conservation. It is equally germane to the design and use patterns of more mundane materials.

Material designers and manufacturers should also be aware of increasing societal sustainability awareness, influencing consumer preferences and willingness to accept products and materials with lower environmental footprints and maintenance costs along with improved technical performance in the use phase. These factors exert market and regulatory pressures on material choice and use, creating an incentive for product manufacturers to innovate to better serve future markets increasingly shaped by sustainability pressures.

4.2.6 The Post-Use Life Cycle Stage

The 'waste hierarchy' is commonly invoked when considering the post-use phase of the life cycle, once products have reached the end of their useful service life. There are various articulations of the waste hierarchy, but one of the most common is to break it down into five phases comprising: (1) prevention; (2) reduction; (3) reuse; (4) recycling; and (5) responsible disposal. Self-evidently, these processes must be addressed in an integrated way, asking up front, for example, whether the product is even needed, or if the service it provides can be met in other potentially less material-intensive ways.

The hierarchy sets an order of priority from a sustainable development perspective. Each layer of the waste hierarchy has its own internal complexities. For example, at the reuse layer, some materials can be recovered for reuse (for example building panels) and the same is also true of entire functional assemblages that can allow products to be reconstituted.

The recycling level of the waste hierarchy is a substantial oversimplification, spanning a wide diversity of approaches to cyclic use with radically different sustainability implications. Taking the diversity of different types of plastic as one example, some can be efficiently mechanically recycled with a relatively low input of energy for melt and extrusion processes generating recyclate of high quality directly enabling cyclic reuse. Conversely, other types of plastic, particularly crosslinked and thermoset polymers, can only be chemically recycled, which entails breaking them down to monomers for reformulation with very substantial energy and chemical inputs. The two approaches have radically different sustainability footprints. Figure 4.2 illustrates how the uses of different materials, particularly in the post-use phase under different reuse or disposal scenarios, can exert significantly different sustainability footprints. Material type matters, including its inherent potential for recycling which, of course, may also be influenced by other constituents added throughout product life including in the compounding and converting as well as use phases. The existence of, or investment to create, recycling infrastructure also matters a great deal, influenced by market forces as discussed later in this chapter.

The existence or development of markets for recyclate is also an important driver for the uptake of recycling. Market forces and social infrastructure, including their supporting investments, cannot be dissociated from material life cycle sustainability if the potential for circular use is to be converted into actual recycling.

Current market forces also exert a significant influence on the actual uptake of recycling. For example, best estimates suggest that around 212,582 tonnes of gold have been mined throughout history, visually presented as forming a cube of pure gold around 22 m on each side, with almost all this gold still in use in one form or another.[7] Across the European Union, 44% of copper demand is met by recycled sources, with 70% of copper in end-of-life products recycled including 90% of that used in civil infrastructure.[8] The overall recycling rate of silver from post-consumer waste globally lies between 30% and 50%.[9] For glass, the UK recycles around 76.5% of all used glass, with each tonne of glass recycled saving around 385,000 tons of CO_2 emissions and averting glass ending up in landfill where it could take up to one million years to degrade.[10] In Germany, the Rewindo programme has contributed to a recycling rate of around 85% for used uPVC window profiles in 2019.[11] Across Europe, a paper recycling rate of 70.5% was achieved in 2022, on track to reach 76% by 2030 enabled by extensive investments by the industry in recycling infrastructure valued at several billion Euros.[12] These major successes rely on material purity, inherent capacity for recycling, investment in recovery and recycling infrastructure, and the value of recyclate for the remanufacture

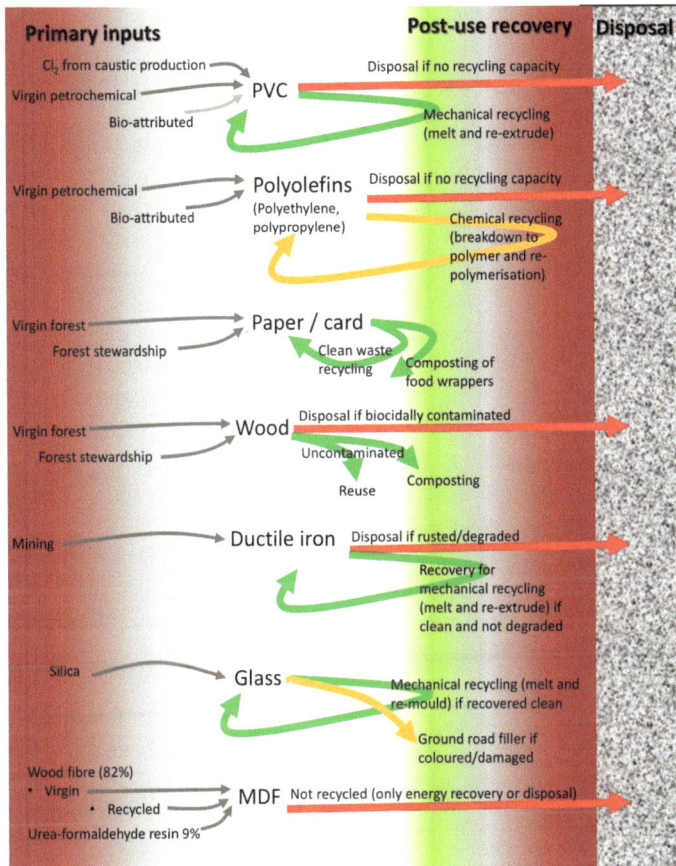

FIGURE 4.2 Illustrative life cycles of different material types with a focus on differences in sustainability footprints across varied forms of post-use recovery or disposal: green arrows signify lower footprints; amber signifies greater impact; and red arrows reflect wastage through material disposal.

of new products. Recycling in the post-use phase ramifies back across whole societal product life cycles, retaining embedded chemical, energetic and economic value in beneficial cycles rather than wasting resources and liberating materials as pollutants.

The term 'responsible disposal' also covers a very wide range of processes from incineration, with or without energy recovery, potentially leading to aerial emissions as well as concentrated pollutants in residual dust, through to landfill with its significant habitat displacement and potential for accumulation and eventual leaching of problematic substances. Composting organic materials enables value recovery if contamination with problem substances is rigorously avoided, whether in the material synthesis, compounding, product use or during potential mixing in the post-use phases.

There is an inevitable need for greater circularity. Substances that cannot be recycled, or that are biocidally treated to prolong use and that therefore finally constitute toxic waste, have to be considered in terms of overall sustainability footprint across product life cycles. Even metals and paper or card, inherently recyclable, may become problematic if compounded by coatings that may inhibit recovery and reuse at the post-use phase.

Significantly, post-use considerations include not only means of disposal but also the inputs and consequences of different types of recycling. This includes their efficiency with respect to energy, chemical and labour inputs, some stemming from established societal resource use practices and others from the characteristics of materials being recycled. These broader considerations are important, and their footprints may potentially far exceed the narrow focus on intrinsic properties of raw or synthesised materials.

The issue of legacy substances can be a confounding factor in the post-use phase of products and materials. Substances permitted historically may subsequently be revealed as of concern. If an absolutist approach is taken to simply banning any contamination of recyclate with these newly classified legacy substances, waste streams containing them may simply be diverted to non-beneficial disposal. As noted in Chapter 2, we will always be revealing new legacy substances, including 'natural' substances often unhelpfully considered 'safe' or benign. Consequently, too inflexible and immediate an interpretation of 'clean chemistry' production applied to recyclate risks condemning ourselves to linear disposal and virgin resource dependence, with all their negative sustainability implications. It is therefore necessary to think more strategically about how to handle inevitable legacy constituents within materials in end-of-life products, particularly for long-life products with a longer history of potential 'legacy' constituents.

A technical press article that I published in 2020,[13] building on another published in 2014, addressed the conflict between immediate expectations of attainment of the goals of the European Commission's *Towards a non-toxic environment strategy*[14] and the equally laudable EC *Circular Economy: Implementation of the Circular Economy Action Plan.*[15] There are inherent tensions between these strategies in terms of resource recovery and the re-entry of 'problem' constituents of end-of-life products entering recycling streams.[16] As a pertinent example, lead was formerly widely used in polyvinyl chloride (PVC) stabiliser additives in Europe, imbuing long-life PVC products with durability and hence long service life per unit of chemical material requiring no maintenance inputs in use. Lead-based additives to virgin

PVC have though been voluntarily substituted across Europe since the end of 2015. However, as these stabilisers imbue PVC products with very long life cycles, particularly in the building sector, lead-containing PVC enters waste streams many years after installation. An overly purist view and expectation of immediate and total attainment of 'clean chemistry' as the strategic aim of the *Towards a non-toxic environment strategy* would inevitably prevent beneficial reuse of this valuable recyclate, thwarting progress with the EC's *Circular Economy: Implementation of the Circular Economy Action Plan* aspiration and dissuading necessary investment in recycling facilities, processes and markets. One has to ask if the dystopian outcome of perpetuating wasteful and polluting linear resource life cycles is really the 'holy grail' of European policy with regard to 'clean chemistry' given that, as we have seen, continued identification of new 'legacy' substances is all but a certainty. Materials in use by society also embody high value: an estimated 150–200 million tonnes of PVC are in beneficial, long-life uses across Europe alone. Is absolute rejection of any beneficial recycling of lead-containing end-of-life materials, even when that lead is firmly immobilised – embedded into the recyclate and unable to leach out into human tissue and the wider environment – really contributing to sustainable development? Whilst lead exposure is not good for people or the environment, a strategic pathway forward can lead incrementally towards the achievement of both goals over a longer-term 'glide path' as 'cleaner' PVC increasingly enters recycling plants. Let's remember that definitions of the term 'strategy' relate to intended directions of travel, not to immediate attainment from the starting point of the deeply unsustainable norms of the suboptimal world we have inherited today. A more intelligent approach is segregation of PVC waste containing a proportion of 'legacy' lead or other substances, which can either be directed towards safe uses of recycled PVC or, better still, blended with 'cleaner' virgin or recycled PVC to ensure that legacy content is below safety standards. Blending of more contaminated with cleaner sources of raw water is a widely accepted approach for meeting quality standards for drinking water, so the principle is established and widely deployed and accepted already. Simplistically, who does not want a 'clean chemistry' environment? Pragmatically, what are the costs of immediate expectation of its attainment in a world that is far from 'clean'? We can, in short, not afford to let an instantaneous expectation of, and leap towards, instant attainment of 'the best' prevent us from innovating and taking 'good', rationally science-based, materially efficient and profitable steps that progressively lead us towards a longer-term sustainable future. And, as virgin materials are manufactured according to current clean chemistry rules, they will eventually enter recycling streams, creating a glide path to progressively cleaner recyclate. In the interim, recycled PVC with legacy substances can be used, potentially with blending, in applications for which they pose no risks. Both strategies are thereby resolved within a practical vision, though certainly not under naive expectations of immediate fulfilment.

It is vital that businesses have the confidence to invest in recycling infrastructure in the post-use phase. Intentions by governments around the world to achieve a circular economy cannot be realised without investment by businesses, in collaboration with other sectors, in the necessary recovery and recycling processes and facilities. Partnerships across sectors, including clear regulatory and fiscal signals as well as facilitation by municipal government bodies and the support of civil society (often vectored through NGOs and the media), are vital for the establishment and success of new recycling businesses. There is often an unrealistic expectation that it is material manufacturers who will take back and recycle these materials when products reach the end of their useful lives. This, though, is not the key skill of manufacturers. The operating environment has to be conducive for the promotion or founding of specialist sorting and recycling companies if we are to realise widely stated aspirations for progress towards a circular economy. It is also vital that viable markets for high-quality recyclate are promoted, for example by using the weight of public procurement prioritising recycled products or content to stimulate the recycling sector (the *Recycled First* policy of Sustainability Victoria in Australia is an exemplar described in Chapter 7) to promote confidence for it to invest. This form of promotion can also stimulate diversification in products containing recycled material, with spill-over promotion into other markets.

4.3 PARADIGMATIC CHANGE FOR SUSTAINABILITY

Arguments about the importance of and urgent needs for sustainable development are so deeply embedded in societal rhetoric that there is barely a need to further justify its necessity for biophysical and economic survival into the longer-term future and to avert potential civil conflict due to competition for depleted resources. Significant issues of equity are raised by our continuing failure to constrain knowingly destructive habits, inevitably precipitating much-impoverished prospects for future generations as well as for less favoured sectors of society in the present. It is worrying that some nations see their own immediate economic empowerment as a sole priority regardless of the associated vandalism of the natural world and its consequences for the future of all.

There are also clearly business imperatives, as dependence on depleted materials or those that we now know to have significant associated risks can lead to uncertain returns on investment as well as potentially mounting future legal liabilities. Above all, we need to select or innovate materials wisely cognisant of their total sustainability footprints wherever their constituents are sourced, rejecting trade with pariah nations or companies ignoring social and environmental responsibilities, to optimally address the needs of a growing human population with rising per capita demands subsisting on a diminished and declining resource base. Innovation in the materials sector is vital to enable society to address its needs more efficiently, safely and responsibly, but also for to secure future profitability better insultated from market shocks and customer deselection.

4.3.1 Joining Up the Life Cycle

From the preceding analyses, it is quite clear that it is essential not only to account for materials at discrete single life cycle stages but also to evaluate the implications of their use from raw material extraction through to integration into finished products and onwards to post-use. Sustainability implications neither stop nor start at the material manufacturer's factory gates. Environmental and ethical issues ramify along the entire value chain, as evidenced by businesses facing consumer, NGO and shareholder challenge due to child labour and other ethical as well as environmental 'skeletons in the cupboard' in their supply chains. There is also now growing scrutiny of carbon budgets as well as ethical practices and other facets of sustainability embedded across entire product life cycle. These ramifications also flow backwards along value chains, for example with large global drink manufacturers facing societal accusations that they are major drivers of the accumulation of plastic litter in the marine environment due to their massive use of single-use packaging; all players in that material life cycle share the risks associated with negative perception.

We have come from a culture in which each phase has been addressed largely in isolation, with risks and responsibilities shed down the life cycle. However, all stages of the cycle are intimately interconnected with each other, and design considerations in any phase from material synthesis right through to capacity for beneficial reuse need to be regarded in an integrated way across societal sectors. Material design considerations therefore need to take account of every stage in integrated value chains regardless of whether the material manufacturer has primary responsibility for that phase. This level of cross-societal joining-up is an indicator of the level of culture change that is implicated in the realisation of sustainable use of materials. Connected thinking across the entire value chain is essential to optimise the sustainable performance of constituent materials, challenging and overturning prevalent perceptions and statutory obligations that divert attention from positive and negative implications affecting all links in the chain. Material manufacturers can help compounders and converters to efficiently make products that are then known to be durable and to require low maintenance inputs and that are then inherently recoverable and recyclable post-use. Novel regulatory approaches are required to drive synergies across product life cycles, realising common benefits and progress towards sustainability.

4.3.2 Consideration of Additives

Additive substances to some materials have received a significant degree of attention from regulators and NGOs. Rightly so, as sustainability considerations relate as directly to chemical additions as to the primary materials themselves. All of these substances have distinct life cycles of their own, so it is important to consider their chemical, physical, human health and broader social impacts from extraction of their raw constituent materials, in manufacture, associated with packaging and distribution, their inclusion into other materials at the compounding phase, when those compounds are converted into final products, onwards in the product use phase, and finally also the implications that arise when products reach the end of their useful lives. Often, as with all materials, scrutiny has largely rested to date on potential hazards relating to the intrinsic chemistry of additive substances. However, this narrow, hazard-based approach overlooks three important facets of the sustainability of additive use.

Firstly, what is the actual degree of risk if the additive substance is wholly consumed in tightly controlled manufacturing processes? Also, where is the risk if the additive is firmly bound in compounds and products, particularly when recycling infrastructure is in place to recover them at product end-of-life averting human and environmental exposure and instead enabling continual beneficial circular use?

Secondly, what functional contributions do the inclusion of these additive substances make towards the meeting of needs, as their purpose is to confer benefits such as product durability and long service life, flexibility or other technical performance attributes? This extends to upstream additive design to facilitate product manufacturing with lower material and energy inputs. The additive's downstream contribution to product service life may also be significant, including extending service life with reduced maintenance inputs, and onwards to compatibility with recycling to promote, or at a minimum not to inhibit, cyclic and beneficial recovery and manufacture.

Thirdly, have we considered all additives across the product life cycle? Impacts of chemicals, energy and labour during the product use phase, be that of biocides to preserve the product or for cleaning or other purposes to maintain serviceability, are just as germane to the overall sustainability footprint as those additive substances that are compounded into the material to modify its performance and contribution to how finished products meet human needs. Additive substances during the product use phase can, as we have seen, potentially become a dominant influence in long-life applications. Furthermore, embedded additives such as brominated flame retardants in polyolefins or fibres, both natural and synthetic, may inhibit the recyclability of products at end-of-life, as may additions such as preservatives and other additives during the use phase of potentially biodegradable materials. Alloys and coatings also constitute additives. All require equally objective evaluation with other materials on a level playing field of sustainability principles.

4.3.3 Joining Up the Parallel Strands of Sustainability

The preceding considerations by life cycle stage also make it clear that there is a need to consider all of the parallel strands of sustainable development in innovation and forward planning.

Conflicts between positive and negative outcomes are common in material use situations, though are missed by too narrow a focus on single issues whilst disregarding collateral implications. The example of substitution of fossil carbon-based materials with biologically based ones exemplifies how burdens can be shifted, in this instance from emissions of climate-forcing gases through to intensification of land use with all its environmental ramifications and the potential dispossession of other land users (including emissions associated with mobilising soil carbon and energy inputs to farming practices). As a further example, cobalt additives may enhance the efficiency and contribute to the durability of some designs of solar panels, enhancing renewable energy generation throughout the use phase of the products. These metals may also be recyclable where suitable recycling infrastructure exists in the regions in which they are installed. However, cobalt is a substance of concern in terms of pollution as well as habitat and aquifer degradation during raw material sourcing, particularly when derived from regions with repressive regimes or where child or indentured labour is either explicitly deployed or implicitly sanctioned. Which is the greater good or ill: decarbonisation in developed world settings where many solar panels are installed, or distributional environmental and social costs incurred in generally poorer cobalt-producing regions? How is this traded off, and are users along the whole value chain not tainted by adverse impacts and potential exposure to publicity even if we choose to turn a blind eye to the inconvenient realities associated with issues that might otherwise get in the way of immediate utility and profitability?

By and large, vested and generally connected interests promulgate and market the benefits, overlooking or suppressing 'inconvenient truths' relating to wider ramifications. Yet all ramifications matter in a world subject to inevitably tightening sustainability pressures and to ever more pervasive digital scrutiny. For example, exposure of the involvement of child labour in some sportswear supply chains fed back directly to the reputation and financial performance of certain brands. The same is true for manufacturers, retailers and other businesses exposed as dependent upon materials imported from repressive regimes or extraction of raw materials associated with damaging pollution. It is clear that all stages along the total societal value chain may be tarred with the same brush if issues such as child labour, damaging strip mining, reliance on unsustainable inputs such as irresponsible palm oil production and other features are overlooked. It is for this reason that there are, for example, standards emerging around substances such as 3TG open (tin, tantalum, tungsten and gold) and some other mined substances, as well as some forest and marine products. All these ramifications affect the value chain right the way through to the use phase.

Taking account of all aspects of the sustainable development framework – chemical, physical, societal and economic – is manifestly essential if we are to avoid simply shifting burdens and risks. Knowledge gaps anywhere along the life cycle can potentially lead to exposures with ramifications for the total value chain. Uncertainties are inevitable, but the aspirational 'gold standard' is that all issues are auditable along the value chain. Auditing of sustainability claims has become mainstream in advanced economies, for example under the ISO 14001 series of standards. Having confidence that there is evidence to back up all aspects of life cycle sustainability pertaining to material usage will be increasingly important for the market as a whole, as will compliance with regulations and societal expectations as sustainability pressures increasingly impinge on customer, consumer and regulatory expectations. In Europe, a Corporate Sustainability Reporting Directive is adding some statutory backing to this, and a similar standard is being introduced in Australia (both considered in more detail later in this book).

4.3.4 Refocusing on the Meeting of Needs

Another significant learning is that it is not just the perpetuation of profit from what we do today that will assure us of sustained business returns and progress towards sustainable development. Rather, continued profitability is best served by refocusing on the needs that these materials and the products that we make from them will serve. Are these needs real or are they frivolous, and will they exist in the future? If these needs do exist in the future, what is the best way of meeting them without an automatic assumption that vested interests in a particular material will offer that optimal solution?

As noted in Chapter 2, in 1987 the world accepted the 'Brundtland definition' of "...*development that meets the needs of the present without compromising the ability of future generations to meet their own needs*". However, transposition of that bold intergenerational commitment to sustainable development into common parlance, regulation and supporting tools has tended to fit into the pre-existing paradigm of 'being less harmful' or of 'treading lighter on the Earth'. The focus on meeting needs now and into an indefinite future, explicit in the 'Brundtland definition', has been lost in translation. Regrettably, this has trapped the aspirations of states and their agencies within the pre-existing resource use paradigm, with suboptimal commitment to eco-efficiency rather than groundbreaking innovation. Using a range of case studies conceptualising how a needs-based approach could proactively complement the focus on establishing acceptable (or at least accepted) baseline performance, I together with my colleague Jim Longhurst addressed the fact that the world at large had failed to engage with this 'missing half' of the sustainable development narrative – the meeting of needs now and tomorrow – presenting different perspectives via a set of case studies.[17] This is of fundamental importance as it is the inspirational aspect of understanding of sustainable development and their implementation that speaks to positive visioning, innovation for connected societal benefit, profitability and opportunity. We must innovate to meet needs, and chemical use is a significant part of how needs are satisfied. Furthermore, it is far more motivating to address material choice and innovation in terms of better meeting human needs on a safe and efficient basis than it is to seek merely to be less harmful!

Regrettably, an anachronistic paradigm still pervades not only common perceptions but also regulatory approaches around the world, modified but not reconstructed from their original focus on 'bad' materials and potential for harm founded on intrinsic chemistry rather than actual material use and benefit delivery. As such, many regulations arrest innovation by focusing almost solely on the potential for hazards and overlooking the stimulation of progressive contributions to better meet needs. This acts as a brake rather than a stimulant of innovation. As we have observed many times, no material is automatically sustainable. Material sustainability has therefore to be assessed with respect to its behaviour in different applications across whole societal life cycles and its contribution to meeting needs safely and efficiently. For example, sandwiches, potato chips or other fast-food wrappers that will inevitably be contaminated with food will generally end up in mixed waste streams, so it makes no sense whatsoever to make them out of durable materials that cannot be benignly composted or otherwise safely reintegrated or combusted. This is an appropriate use of paper and other naturally derived materials. By contrast, long-life applications, such as many building products, can be best served by durable polymer, glass, metal, plastics, ceramics or other materials that are resilient and offer long service life per unit of material resource and require little or no maintenance inputs. Optimally, they should then be recovered for mechanical recycling at product end-of-life. Wise selection and deselection can only be determined when contextualised in whole-life sustainability assessment, taking account of social, chemical, occupational exposure and physical attributes in an integrated way.

4.3.5 Opportunities for the Circular Economy

In the paper industry, there is recognition that the 'urban forest' of paper and card in circulation in society constitutes an alternative source of wood fibre to extraction from natural or managed forests, substantially reducing energy and material consumption whilst averting waste.[18]

There is also growing interest in the recovery of materials formerly disposed of in landfill sites, retrospectively converting waste materials into new resources. This process, known as landfill mining and reclamation, entails excavation and processing of buried solid waste matter to recover potentially valuable recyclable materials such as scrap metal, combustible fractions that may be used for power generation, as well as soil formation. Residual waste is replaced at a lower volume into the landfill site when usable constituents have been recovered, saving space particularly where landfill capacity is limited.

As we saw with the high rate of recycling of gold, copper and silver, paper and glass – at least in Europe – material purity, the political environment, investment in infrastructure and the value assigned to recycled materials are significant enablers of circular use. Today, artificially low raw resource prices and both structural and economic obstacles mean that recycling of plastic, amongst other materials, struggles to be profitable particularly in the US following China's refusal in 2018 to import the world's waste.[19] A renewed intent announced in the US in 2025 to remove the perceived constraints of environmental responsibility and concern for climate change to drilling for new oil reserves will also further depress the costs of virgin supply, should companies and other nations choose to continue to trade with what in objective terms has become a pariah regime rejecting responsibilities formerly agreed at intergovernmental level.

Investment in and prioritisation of recycling post-use is a society-wide issue, requiring joined-up investment. Sorting of spent products and constituent materials is a matter for citizens as well as all sectors of society, including waste handlers as well as those permitting recycling facilities. Recycling cannot simply be seen as the responsibility of material producers, with others who benefit from material use affecting the use cycle along the value chain abrogating responsibility. Furthermore, material manufacturers are not ideally equipped to also become recyclers, as recycling is a societal concern requiring the development of a dedicated recovery and recycling sector and associated markets for recyclate. Cross-societal capacity-building is an absolute requirement if progress is to be made towards meeting aspirations for circularity and sustainability.

4.4 STILL SEEKING A LEVEL PLAYING FIELD

My 2024 book *Seeking Sustainable Development on a Level Playing Field*[20] introduced the importance of a consistent and generic approach to thinking about societal choices, innovation and regulation of material use informed by sustainability principles. Aspects of this have been outlined already: risk over the whole societal life cycles of the products into which materials are incorporated; consideration of chemical, physical and socio-economic aspects of sustainable development as a systemically connected whole; recognising and rewarding positive contributions towards serving needs; challenging assumptions that what we make today will continue to best serve that purpose; a backcasting approach that recognises that the conditions, markets and opportunities of the future will inevitably be substantially modified by tightening sustainability pressures; and integration of players along value chains. Collectively, these represent paradigmatic changes rather than simple adjustments to today's norms. These changes, though, pre-empt pressures and changes in markets, offering continuing profitability if informed by opportunities presented by the conflict of rising human demands with dwindling and increasingly contested natural capacities. Collaborative, vision-led working across societal sectors is a necessary aspect of the transition of global society consistent with often-repeated rhetorical commitments to sustainable development made over the past four decades.

This book progresses recognition that the disparate elements of society – all of them highly interconnected strands within the complex socio-economic whole through which materials flow – need to act as an integrated whole to achieve security and opportunity through a cultural transformation that is symbiotic with the wider supporting socio-ecological system upon which humanity entirely depends. This is a difficult and necessarily long journey, as the pursuit of a genuinely sustainable pathway of development unavoidably challenges the flawed paradigms and vested interests widely and deeply entrenched in today's norms. We need instead to be guided by how best to address the meeting of needs in a more resource-limited future, setting aside presumptions and prejudices relating to materials in current use and naive judgements about whether they are inherently 'good' or 'bad' or what their intrinsic properties imply when addressed in isolation from real-world risk. The level playing field approach is inherently 'material blind', acknowledging that different materials may be more or less appropriate for servicing needs in different application types.

However, there is more. To attain this, we need to look beyond technical questions, recognising the vital importance of attaining greater symbiosis between currently often fragmented societal sectors to harness their collective vision and energies to make real and material changes in cultural habits and aspirations.

NOTES

1 Responsible Minerals Initiative. (n.d.) *Conflict Minerals Reporting Template.* Responsible Minerals Initiative. [Online.] https://www.responsiblemineralsinitiative.org/reporting-templates/cmrt/, accessed 12 October 2024.
2 FSC United Kingdom. (2023). *What is FSC?* Forest Stewardship Council UK. [Online.] https://uk.fsc.org/what-is-fsc, accessed 20 April 2024.
3 RTRS. (2024). *What are the Benefits of RTRS Certification?* Round Table on Responsible Soy (RTRS). [Online.] https://responsiblesoy.org/certificacion?lang=en, accessed 23 September 2024.
4 WWF. (2014). Handle with care: understanding the hidden environmental costs of cotton. *World Wildlife Magazine*, Spring 2014. [Online.] https://www.worldwildlife.org/magazine/issues/spring-2014/articles/handle-with-care, accessed 12 October 2024.
5 Better Cotton. (2024). *What is Better Cotton?.* Better Cotton. [Online.] https://bettercotton.org/, accessed 12 October 2024.
6 WWF. (2014). Handle with care: understanding the hidden environmental costs of cotton. *World Wildlife Magazine*, Spring 2014. [Online.] https://www.worldwildlife.org/magazine/issues/spring-2014/articles/handle-with-care, accessed 12 October 2024.
7 World Gold Council. (2024). *Above-Ground Stock.* World Gold Council. [Online.] https://www.gold.org/goldhub/data/how-much-gold, accessed 12 October 2024.
8 EuRIC. (n.d.) *Metal Recycling Factsheet.* EuRIC AISBL – Recycling: Bridging Circular Economy & Climate Policy. [Online.] https://circulareconomy.europa.eu/platform/sites/default/files/euric_metal_recycling_factsheet.pdf, accessed 13 October 2024.
9 REMONDIS Recycled Raw Materials. (2024). *Working Hard at Recycling is Well Worth Its While.* REMONDIS Recycled Raw Materials. [Online.] https://www.recyclingrohstoffe.de/en/material-streams-climate-change-mitigation/technology-metals/, accessed 16 October 2024.
10 Businesswaste.co.uk. (2024). *Glass Waste Facts and Statistics.* [Online.] https://www.businesswaste.co.uk/your-waste/glass-recycling/glass-waste-facts-and-statistics/, accessed 13 October 2024.

11 Schüco. (2021). *Sustainable Building with Windows Made from Recycled Plastic.* Schüco Polymer Technologies. [Online.] https://www.schueco.com/de-en/home-owners/inspiration/sustainable-building-with-windows-made-from-recycled-plastic, accessed 13 October 2024.

12 EPRC. (2023). *Monitoring Report 2022: European Declaration On Paper Recycling 2021–2030.* European Paper Recycling Council (EPRC). [Online.] https://www.cepi.org/wp-content/uploads/2023/09/EPRC-Monitoring-Report-2022_Final.pdf, accessed 13 October 2024.

13 Everard, M. (2020). A lead on recycling PVC. *IOM3 Features* (substitute for Materials World magazine during Covid-19 shut-down), https://www.iom3.org/resource/a-lead-on-recycling-pvc.html, accessed 25 December 2024.

14 EC. (2017). *Towards a Non-Toxic Environment Strategy.* European Commission, Brussels. http://ec.europa.eu/environment/chemicals/non-toxic/index_en.htm.

15 EC. (2018). *Circular Economy: Implementation of the Circular Economy Action Plan.* European Commission, Brussels. http://ec.europa.eu/environment/circular-economy/index_en.htm.

16 Smith, N.C. and Jarisch, D. (2016). INEOS ChlorVinyls: a positive vision for PVC (A). In: Lenssen, G.G., Smith, N.C. (eds), *Managing Sustainable Business*, pp. 73–106. Springer, Dordrecht.

17 Everard, M. and Longhurst, J.W.S. (2018). Reasserting the primacy of human needs to reclaim the 'lost half' of sustainable development. *Science of the Total Environment*, 621, pp. 1243–1254. https://doi.org/10.1016/j.scitotenv.2017.10.104.

18 For example, Kate, L. (2018). Your 'urban forests' and why they're important for recycled paper supplies. *Triple Pundit*, 01 August 2018. [Online.] https://www.triplepundit.com/story/2018/your-urban-forests-and-why-theyre-impor-tant-recycled-paper-supplies/11286, accessed 13 October 2024.

19 Joyce, C. (2019). *U.S. Recycling Industry is Struggling to Figure Out a Future without China.* NPR.org, 20 August 2019. [Online.] https://www.npr.org/2019/08/20/750864036/u-s-recycling-industry-is-struggling-to-figure-out-a-future-with-out-china, accessed 25 December 2024.

20 Everard, M. (2024). *Seeking Sustainable Development on a Level Playing Field: A PVC Case Study.* CRC Press, Boca Raton, FL.

Regulation as Enabler or Barrier to Sustainable Development

5

Statutory regulation of chemical manufacturing, use, disposal and pollution as well as import and export is accepted as a norm today. However, its genesis has a long history, is still evolving and has variable relationships with the stimulation of sustainable use of materials. Regulation of chemical manufacture, marketing, use, translocation and disposal is complex but has wide ramifications for protecting public health, the environment and economic development. Governments around the world have acted upon a range of differing philosophical approaches to regulation, reflecting cultural, political and economic differences, and have instituted an assortment of enforcement bodies.

Questions need to be raised regarding whether our current regulatory regimes are fit for the purpose of driving the innovations necessary for sustainable development, including the sustainable use of materials. If not, what are the necessary features of an appropriate regulatory model?

5.1 A BRIEF HISTORY OF CHEMICAL REGULATION

A long history of regulation pertains to the release of potentially harmful materials into the environment. As far back as the 1200s, royal edicts in Britain banned the dumping of animal waste into watercourses with

DOI: 10.1201/9781003637875-5

stringent punishments for offenders. However, the world's first formal industrial regulation was implemented in 1863 in the form of the Alkali Act, put in place by the Parliament of the UK. The initial purpose of the Alkali Act was to manage the aerial discharge of muriatic acid (gaseous hydrochloric acid) from alkali works. This foundational Act was subsequently amended and extended many times to address other polluting substances of concern discharged by industry. From the 1920s, the focus shifted solely from heavy industry towards wider industrial emissions. A sequence of reformed Alkali Acts was replaced in the UK by the Environmental Protection Act 1990.

Various European nations have similar histories of evolving environmental legislation and associated management authorities. The European Union is the source of a range of environmental Directives unifying approaches across the bloc. Early examples include the 1967 Dangerous Substances Directive (*Council Directive 67/548/EEC of 27 June 1967 on the approximation of laws, regulations and administrative provisions relating to the classification, packaging and labelling of dangerous substances*). Also, the 1978 Freshwater Fisheries Directive (*Council Directive 78/659/EEC on the quality of fresh waters needing protection or improvement in order to support fish life*) specified general water quality standards including lists of substances of concern, though this Directive has since been superseded and repealed. Other environmental Directives pertaining to the control of environmental releases of substances were to follow, such as the *Urban Waste Water Treatment Directive* in 1991. Many other subsequent Directives list substances of concern requiring control, substitution or strict use criteria. A small subset illustrating their range is listed in Box 5.1. European Member States are required to transpose these and more Directives into domestic legislation. Significant amongst these is the REACH (Registration, Evaluation, Authorisation and Restriction of Chemicals) regulation that entered into force in June 2007, aimed to "...*improve the protection of human health and the environment from the risks that can be posed by chemicals, while enhancing the competitiveness of the EU chemicals industry*".[1] REACH also promotes alternative methods for the hazard assessment of substances, which include not only those used in industrial processes but also in day-to-day lives such as in cleaning products, paints, clothing, furniture and electrical appliances.

BOX 5.1:   SUBSET OF EUROPEAN UNION DIRECTIVES LISTING SUBSTANCES OF CONCERN REQUIRING MANAGEMENT

- *Regulation 450/2009/EC – Food Contact Active and Intelligent Materials – CMR Substances Not Allowed for Use*, which establishes specific requirements for the marketing of active and intelligent materials and articles intended to come into contact with food.

- *Regulation 1107/2009/EC – Placing of plant protection products (PPP) on the market.*
- *Regulation 1223/2009/EC – Cosmetic Products Regulation, Annex II – Prohibited Substances.*
- *Regulation 10/2011/EU – Plastics Food Contact and Articles Regulation,* Annex I of which list 'Authorised Substances' with the prohibition of substances not included in this list for use in the production of plastics materials and articles intended to come into contact with food.
- *Regulation 66/2010/EC – Ecolabel – Restrictions for Hazardous Substances/Mixtures,* Article 6(6) of which bans the issuance of an ecolabel to goods containing substances or mixtures classified as toxic, hazardous to the environment, CMRs (carcinogenic, mutagenic or reprotoxic), or containing SVHCs (Substances of Very High Concern as classified under Article 57 of the EU REACH regulation).
- *Regulation 305/2011/EU – Construction Product Regulation,* various articles of which require SDSs (safety data sheets) and stipulate that construction works must not have a high impact on human health or the environment due to emissions.
- *Regulation 2017/745/EU – Medical Devices Regulation – Hazardous Substances,* which lays down rules concerning the marketing or use of medical devices containing hazardous substances.
- *Directive 89/391/EEC – Occupational Safety and Health (OSH) Framework Directive – Hazardous Substances,* introducing measures to encourage improvements in the safety and health of workers including risks arising from chemical, physical and biological agents at the workplace.
- *Directive 2000/53/EC – End-of-Life Vehicles Directive – Hazardous Substances,* mandating prevention of waste from vehicles and the reuse, recovery and recycling of end-of-life vehicles and their components.
- *Directive 2004/37/EC – Carcinogenic, Mutagenic and Reprotoxic (CMR) Directive, Annex I – Substances, mixtures, and processes,* setting minimum requirements for protecting workers against risks to their health and safety arising, or likely to arise, from exposure to CMR substances at work.

Instigation of the Environmental Protection Agency (EPA) in the US in 1970 has already been addressed in Chapter 2. The EPA was initially charged with implementation of the US Clean Air Act (1970) but its remit has continued to evolve, for example to include authorisation and regulation of pesticides, water quality, waste sites, climate change and controls on the use of chemicals. The US Toxic Substances Control Act (TSCA),[2] passed in 1976, is administered by the EPA. The TSCA is one of the most significant strands of the regulation of chemical substances not already covered by other federal US law, spanning chemicals already in commercial use as well as novel substances. One weakness is that pre-existing chemicals in use before the TSCA was put into place continue to be permitted for use by default and, hence, are implicitly authorised under a 'grandfathering' approach. However, the TSCA can also be applied retrospectively to regulate chemicals in use prior to 1976 if they are perceived to pose "*...unreasonable risk of injury to health or the environment*" based on a risk assessment approach. Additional US chemical regulations include the Consumer Product Safety Improvement Act (CPSIA) of 2008. The CPSIA imposes new testing and documentation requirements and new acceptable levels of several substances on manufacturers of a range of products (apparel, shoes, personal care products, accessories and jewellery, home furnishings, bedding, toys, electronics and video games, books, school supplies, educational materials and science kits).

Many other global regions have similar or, in some cases, differing approaches to the regulation of chemicals. Some examples from different continents are summarised in Box 5.2.

BOX 5.2:   REGULATION OF CHEMICALS IN OTHER SELECTED GLOBAL REGIONS

In Australia, the regulation of chemicals is overseen by a range of bodies charged with specific responsibilities.

- The primary Australian Pesticides and Veterinary Medicines Authority (APVMA) has responsibility for pesticides and veterinary medicines.
- The Therapeutic Goods Administration (TGA) is responsible for therapeutic goods including medicines, medical devices and blood products.
- The Department of Agriculture, Fisheries and Forestry (DAFF) regulates fertiliser and biocidal chemicals used in agriculture, including fertilisers, herbicides and fungicides.

- The Department of Climate Change, Energy, Environment and Water (DCCEEW) regulates industrial chemicals that pose a risk to the environment, and WorkSafe Australia regulates chemicals used in the workplace. The Department of Health and Aged Care (DHAC) is responsible for the Australian Industrial Chemicals Introduction Scheme (AICIA) assessing the risks of imported or manufactured chemicals for which registration is mandatory.

In Brazil, several government agencies oversee a comprehensive chemical regulatory framework.

- The National Health Surveillance Agency (Agência Nacional de Vigilância Sanitária – Anvisa: ANVISA) is responsible for regulating all types of chemicals.
- The Brazilian Environment Agency (Instituto Brasileiro do Meio Ambiente e dos Recursos Naturais Renováveis: IBAMA) regulates chemicals posing risks to the environment.
- The Ministry of Labour and Employment (Ministério do Trabalho e Empreg: MTE) regulates chemicals used in the workplace.
- The National Institute of Metrology, Standardization and Industrial Quality (INMETRO) regulates chemicals used in industrial processes.

China has established a complex network of government agencies with specific responsibilities for the regulation of chemicals.

- The National Health Commission (NHC) is responsible for regulating pharmaceuticals, medical devices and other health-related products.
- The State Administration for Market Regulation (SAMR) addresses general consumer products including food additives, cosmetics and household chemicals.
- The Ministry of Ecology and Environment (MEE) regulates chemicals posing environmental risks.
- The Ministry of Industry and Information Technology (MIIT) regulates chemicals used in industrial processes.
- The State Administration for Grain Work (SAGW) regulates chemicals used in agriculture.

Other examples amongst many from around the world include:

- The 2016 Canada Consumer Product Safety Act (CPSA) and the Canadian Food and Drug Regulations (C.R.C., c.870).
- South Korea has a *Toxic Chemicals and Restricted Chemicals* classification in its *Substances Subject to Intensive Control List.*
- Japan *Toy Safety Standard* imposes limits on specific substances in toy applications.

5.2 PHILOSOPHICAL APPROACHES TO CHEMICAL REGULATION

Early instigation of regulation of chemicals was largely reactive to the emergence of human health, environmental and nuisance impacts of different substances. Different philosophies have since emerged to underpin regulatory approaches across the world.

A common approach is the precautionary principle (PP), under which uncertainty about potential risks invokes precautionary measures as a basis for protecting human and environmental health. The European Union has established the PP as a cornerstone of its chemical policy and regulatory framework, as for example observed in the REACH regulation governing the perceived safety of chemicals placed on the market. Australia also bases its approach to chemical regulation on the PP, only allowing onto the market chemicals that have been adequately assessed and found to be safe. Brazil too bases its approach on the PP, requiring that precautionary measures are taken to protect human health and the environment when the potential risks associated with a chemical are uncertain.

Contrasting with the PP, a more risk-based approach to chemical regulation is observed in the US. The risk-based approach seeks to identify and quantify risks associated with chemicals as a basis for appropriate mitigation measures.

Between these different approaches, regulatory bodies also commonly adopt targeted strategies to protect vulnerable members of society, such as children or pregnant women or where longer-term health effects may accrue. China deploys a combination of the PP and the risk-based approach to chemical regulation, balancing the imposition of precautionary measures in the face of uncertainty with risk assessment to identify and evaluate the likelihood and severity of exposure to chemical hazards.

Liabilities for the consequences of chemical use are reflected in the 'polluter pays principle' (PPP), assigning responsibility to those who produce or use chemicals. The PPP is widely adopted around the world, serving as a pre-emptive incentive for companies to develop safer products with lower environmental and health impacts.

Another philosophical approach is that of utilitarianism. This approach is focused on the maximisation of overall wellbeing. Applied to chemical regulation, it prioritises benefits for factors such as improved quality of life and economic growth balanced against potential harm. Under the utilitarian philosophy, benefits for agricultural productivity may be balanced with risks associated with the use of pesticides. In medicine, the approval and administration of drugs with known contraindications is determined on the basis of patient benefits weighed against potential adverse health outcomes. In effect, exemptions for specific uses of otherwise banned substances fall under this philosophy. However, the bulk of chemical regulation tends to focus on the potential for negative impacts rather than contributions to beneficial outcomes for society.

An advanced approach to precaution is embodied by The Natural Step (TNS), an international NGO promoting sustainable development founded on a science-based model constructed from basic and non-contentious scientific principles underpinning the dynamics of the planetary ecosystem. From the TNS science model, four 'System Conditions' arise as necessary conditions to be met for sustainability. (These four TNS System Conditions relate to the systematic accumulation of lithospheric substances, accumulation of synthetic substances, physical degradation of supporting ecosystems and structural obstacles to people meeting their needs.) Through this lens, the TNS approach considers systematic increases in the concentration of both lithospheric and human-produced substances in the environment, regardless of known or suspected effects. If systematic chemical accumulation beyond natural background concentrations occurs, there is potential for them reaching thresholds with the potential to trigger formerly unsuspected adverse environmental and/or health outcomes. Historic examples of such thresholds triggering formerly unsuspected negative consequences include the accumulation of carbon dioxide in the atmosphere driving climate change, discovery of endocrine-disrupting and teratogenic effects of a wide range of synthetic substances, and the carcinogenic implications of dust from asbestos, wood and, potentially, such other purportedly inert substances as titanium dioxide.

There is a further interesting construct linking public attitudes with perceived though not necessarily scientifically grounded chemical properties, summed up as 'risk = hazard + outrage'.[3] The weight of public opinion can significantly influence media and wider perceptions and drive regulatory reaction and business deselection even in the absence of robustly grounded evidence.

This spectrum of regulatory and management philosophies – precaution based on hazard, risk assessment, identification of vulnerable groups, potential liabilities, utilitarianism, averting systemic accumulation that may lead to unforeseen threshold effects, and 'outrage' – is also augmented by public, NGO, shareholder and wider perceptions bearing down as an informal regulatory environment to which markets and the statutory sector may respond. Beyond regulatory frameworks, there are strategy documents that may prioritise some materials and discourage the use of others, possibly also backed up by fiscal measures such as taxes and subsidies. Public procurement policies can also be used by governments to leverage their purchasing power to promote material selection or deselection.

5.3 SUPPORTING CHEMICAL ASSESSMENT TOOLS

To enact the intent of regulations, it is obviously necessary to develop appropriate and robust tools and approaches for chemical assessment. This is essential for clear communication, credible certification, as well as transparency and ultimately accountability along the value chain. Consequently, a wide variety of chemical assessment approaches has been put in place, ranging from those based on standards defined by ecotoxicological testing to which safety factors relevant to the chemistry of the substance have been applied, through to more nuanced approaches.

As observed in the previous chapters, taking a whole product life cycle approach is essential if potential hazards relating to the ecotoxicological, teratogenic, ozone-depleting, eutrophicating and other intrinsic properties of materials are to be translated into real-life risks. Many established approaches though are based on simplistic judgements about the potential for negative outcomes, overlooking actual risks as well as, importantly when considering the 'missing half' of sustainable development, benefits that can flow from the use of the substance under investigation. Another limited focus that is commonly applied is a 'cradle to gate' approach without necessarily considering later connected life cycle stages, or at least delegating concerns to other actors downstream in the value chain or to regulators. This leads to poor join-up along value chains and the flow of materials through society, and to choices and innovations that may not best serve the sustainable use of materials and the safe and efficient meeting of needs. Some examples of chemical assessment tools and rating approaches used regionally or globally are listed in Box 5.3.

BOX 5.3: EXAMPLES OF CHEMICAL ASSESSMENT OR CERTIFICATION APPROACHES USED REGIONALLY OR GLOBALLY

- The current mainstream chemical assessment approach applied across the European Union is the REACH regulation, intended to "...*improve the protection of human health and the environment from the risks that can be posed by chemicals, while enhancing the competitiveness of the EU chemicals industry*"[4] through sequential steps reflected in the title – Registration, Evaluation, Authorisation and Restriction of Chemicals – REACH in industrial processes and domestic applications. As discussed in Chapter 2, criteria used for assessment are based on the 'intrinsic properties' of substances or, in other words, consideration of hazard outside of the context of their use and hence actual risk.

- Environmental Product Declarations (EPD) and Product Environmental Footprint (PEF) approaches are also promoted in Europe to report on the total environmental impact of products and substances. Again, these are based largely on the potential for harm using on metrics generally applied in life cycle assessment.

- EcoVadis is a commercial service with global reach and uptake by a number of large corporations, providing sustainability ratings for companies covering a broad range of non-financial management systems including environmental, labour and human rights, ethics and sustainable procurement impacts, presenting evidence-based assessments as scorecards (0–100 scores) and medals (platinum, bronze, silver, gold).[5]

- LEED (Leadership in Energy and Environmental Design) is a building rating system that is US-based but has a global reach, providing "...*a framework for healthy, highly efficient, and cost-saving green buildings, which offer environmental, social and governance benefits*".[6]

- The Green Building Initiative (GBI), an international non-profit organisation and developer of American National Standards Institute (ANSI) accredited standards, was founded in 2004 and is dedicated to improving the impact of the built environment on climate and society. GBI is also the global provider of the Green Globes® and federal Guiding Principles Compliance building certification and assessment programs.[7]

- BREEAM®, founded in 1990, is a globally applied sustainability assessment method for the built environment and infrastructure with third-party certified standards aimed to improve asset performance from design through to the construction, use and refurbishment of buildings.[8]

- Green Star, launched by the Green Building Council of Australia in 2003, is a sustainability rating tool used for buildings, fit-outs and community developments aiming to reward products with: lower environmental impacts; reduced impacts of climate change; enhanced health and quality of life; restored and protected biodiversity and ecosystems; improved resilience in buildings, fit-outs and communities; and positive contributions to market transformation and a sustainable economy.[9]

5.4 CHEMICALS AND OTHER SUBSTANCES

As observed previously, different substances with ostensibly similar hazard profiles do tend to be handled somewhat differently in society. Take, as examples already given, the three IARC Category 1 (proven human carcinogen) substances including wood dust, ozone and vinyl chloride monomer (VCM). Wood dust is not stringently regulated in most sawmills and workshops. Ozone releases are controlled but the substance is in wide use for sterilisation of water, including in municipal swimming pools and also in less closely regulated applications such as in the aquarium hobbyist trade. By contrast, VCM is stringently monitored and controlled with supporting regulation and, at least in developed countries, is now essentially contained in manufacturing to a level where exposure and hence risk are eliminated with the substance then wholly consumed during polymerisation in the production of PVC. There is a lack of consistency in regulation and common understanding relating to the use of these substances of high potential concern.

As observed when listing some of the wide variety of regulatory bodies across the world and within nations, drugs, agrochemicals, veterinary medicines, foods and other chemical materials may be subject to a range of different regulatory regimes and subject to varied degrees of scrutiny depending on application. This contrasts, for example, with stringent controls imposed on the chemical industry and its products. In some instances, this is legitimate, as for example, getting the balance right between potential benefits and potential harm in the application of medicines as well as targeted use of pesticides. However, there are internal contradictions.

One of these contradictions is seen in the case of 'flex crops'. Flex crops are those that can be diverted into different markets – food, biofuel, chemical feedstock or other – depending on market forces. Their increasingly wide uptake has altered production patterns as well as power relations between landholders, agricultural labourers, crop exporters, processors and traders, incentivising changes in land-tenure arrangements.[10] The rise of flex crops also further suppresses agrodiversity, with associated ecological and social impacts. Outputs from these crops are then highly processed into novel forms that may be used, if the market forces favour it, as cheap foods (or 'food-like substances') or food additives such as fats and sweeteners, emulsifiers, stabilisers, colourings and flavourings with dubious health benefits.[11] Alternatively, if the market favours it, these same transformed substances can be diverted into wider industrial, cosmetic, fuel and pharmaceutical applications. This enables farming enterprises to switch production or sales to serve the most remunerative market. In turn, this may drive various forms of inequality, including wealth concentration among super-rich sectors of society,[12] and it has now become a significant driver of global agri-food system change potentially undermining food security.[13] Materials derived from flex crops though may be subject to different regulatory approaches under the purview of different enforcement bodies depending upon whether they are deemed industrial chemicals, fuel, food or food additives, pharmed drugs, or if they serve other purposes. There is, therefore, a lack of consistency and transparency in how these identical or similar molecules are handled.

5.5 INTERGOVERNMENTAL CONVENTIONS

In addition to national and regional regulations pertaining to chemicals, a raft of intergovernmental conventions has also been implemented. These are important legal instruments to ensure a degree of consistent global action around issues of concern, including their translation into transborder trade. Some examples pertinent to chemical management include the Basel Convention (the *Basel Convention on the Control of Transboundary Movements of Hazardous Wastes and Their Disposal*) that entered force in 1992, the Rotterdam Convention (the *Rotterdam Convention on the Prior Informed Consent Procedure for Certain Hazardous Chemicals and Pesticides in International Trade*) concluded in 1998, and also the Minamata Convention (relating to mercury) and the Stockholm Convention (addressing persistent organic pollutants: POPs) both of which are outlined in Chapter 2.

The updating of the requirements, interpretation and implementation of intergovernmental Conventions in response to changing global situations, emerging knowledge and novel strategies is mixed. There are both progressive examples, but also others that have yet to be reinterpreted to meet their foundational best intentions in a much-changed world.

One of the progressive examples of an intergovernmental convention that has moved with the times is the Ramsar Convention (the 'Convention on Wetlands' as also described in Chapter 2), which has been regularly revised through successive Conferences of Parties to include consideration of the ecosystem services produced by wetlands, assessment techniques to achieve this, integration with the Sendai Framework for Disaster Risk Reduction,[14] support of sustainable livelihoods and as educational resources, amongst other novel approaches.

Contrasted with this is, for example, the Basel Convention. Whilst the intent of this Convention is the laudable and necessary control of flows of hazardous waste materials into countries and regions unable to process them, and therefore where hazard might translate more directly into risk, there have been numerous substantial changes in the global situation in the decades since the Basel Convention came into force in 1992. Among these is awareness that a risk-based approach is more insightful than consideration of the potential for hazard alone, careful handling of materials avoiding exposure to scheduled materials meaning that management can control risks to human or environmental health. Also, a strand within the 'Brundtland Report' is the recognition of the importance of developing 'industrial ecosystems', wherein by-products from one industry can be beneficially exploited as valuable resources by other industries, ideally proximally but potentially crossing borders. Escalating trade between borders due to globalisation is another of the substantial changes observed since 1992. Then there is the near-global pivot of resource use strategies away from linear towards commitments to greater circularity, for which recovery and recycling are essential underpinnings. A combination of greater globalised trade flows and commitments to circular use impinge most significantly on island and small nations, small economies or those with scattered population centres making localised recycling less feasible or impossible and therefore necessitating transfer of by-products, spent and post-industrial materials across boundaries for beneficial reuse at scale. This combination of factors means that substances classified as 'waste' under the Basel Convention have to be handled differently from those recognised as by-products feeding into different value chains, and those that fall under a coarse screening as 'hazardous' may be condemned to wasteful disposal if addressed purely as hazards rather than through risk assessment. Where circular reuse can be achieved with assured safe handling of materials crossing borders, denial of this opportunity drives perverse outcomes and wastage of valuable resources. Overly stringent literal enforcement of an unreconstructed 1992 set of words surely works against the best intentions of the Basel Convention.

Reinterpretation of the intent of all international agreements in the light of societal change, and particularly how they can contribute to rather than inhibit progress towards increasingly more urgent, clearer and better-understood sustainability goals, is a priority for regulators and governments. Without periodic revision, aspirations to achieve the sustainable use of materials and wider sustainability goals risk being undermined by narrow interpretation, over-zealous enforcement and failure to resolve conflicts between strategies. This parallels the conflicts entailed by immediate expectation of attainment of the goals of the EU's 'circular economy' and 'clean chemistry' strategies, as described in Chapter 4.

5.6 IS OUR LEGACY REGULATORY APPROACH FIT FOR PURPOSE?

Regulation of chemicals is welcome to eliminate or better manage the most hazardous substances for which exposure during their life cycle can pose genuine risks. However, many pre-existing tools and regulatory approaches now suffer from two principal shortfalls in terms of stimulating sustainable progress.

The first of these is that they essentially focus on potential hazard based on intrinsic chemistry. Hazard is, of course, an important consideration, but it does have to be set in the broader context of actual risk in product life cycles. Avoidance of unintended exposure mitigates actual risk, as previously discussed in relation to hydrochloric acid production and resorption within the closed confines of the human stomach, capture of Category 1 carcinogenic wood dust, the complete consumption of toxic monomers in tightly contained manufacturing facilities, or copper that, though inherently toxic, is recovered in closed loops for reuse at end-of-life of plumbing and wiring applications. If our judgement rests solely on potential hazard based on intrinsic chemistry, we are, in essence, still stuck in the myopia of 'good' versus 'bad' materials regardless of context, actual life cycle risk and whether they can make safe and efficient positive contributions to the meeting of needs.

A second and fundamental shortfall is a misreading of the purpose of sustainable development. As discussed in Chapter 2, the world signed up to bold intergenerational commitments under the 1987 'Brundtland definition', "...*development that meets the needs of the present without compromising the ability of future generations to meet their own needs*". This is a progressive definition, focused on positive outcomes: the meeting of needs now and tomorrow. By contrast, the framing of the vast bulk of regulation tends to put a brake on rather than to stimulate innovation through a predominant focus on being 'less bad' rather than asking searching questions about the most efficient and safe means to meet needs. Negative impacts of course matter, but the bulk of regulations currently fails to engage with this 'missing half' of the sustainable development narrative,[15] perpetuating distortion of the ideal of sustainable development by disregarding it as a forward-looking and progressive process focused on the longer-term attainment of sustainability with many benefits to society and to far-sighted businesses.

If we hamper ourselves by too narrowly focusing on yesterday's or today's emerging or known potential problems, we will most likely continue to lurch from one problem into others we have not yet foreseen. Regrettable substitutions are widespread, for example in the cascade of novel pesticides replacing ones for which hazard has become evident, swapping yesterday's new 'wonder chemical' with others as its problems manifest. An overly narrow focus has also led us, for example, to address energy efficiency without foresight, often locking investment into processes or products that may themselves not have a place in the future. Equally, a lurch away from utilisation of fossil carbon towards biologically based energy and feedstock might, without foresight, inadvertently magnify pressures on land use including habitat conversion, soil erosion, water over-abstraction or diversion from other critical uses, and/ or the disenfranchisement of communities formerly dependent on the habitat or locally nuanced land use as it is converted for intensive agriculture. What is missing is a far-sighted perspective anticipating and stimulating a creative approach to how best to meet needs in future markets that will inevitably be substantially shaped by sustainability pressures.

Measured against these principles, the bulk of current regulatory machinery is not fit for purpose. However, as we will address later in this book, some novel approaches are emerging. It is clear though that it is not just peripheral modification that is required, but rather a paradigmatic change that refocuses attention and innovation on positive, safe and efficient means to meet needs.

There is also a significant degree of inconsistency between nations and political regions, erecting barriers to international cooperation in the form of global treaties that do not evolve with time, insufficient knowledge sharing and implementation of trade agreements or barriers. This is highly significant in a globalised world where trade beneath statutory volumetric thresholds can result in international sales of less safe substances and products, undermining markets in which sustainability considerations are taken more seriously and invested more heavily. If cost remains the principal driver in international trade, unsustainability by the most neglectful state and corporate suppliers will most likely continue to be rewarded no matter how loud the rhetoric to the contrary.

Beyond the predominant focus on the potential for adverse consequences from chemical use, which absolutely matters if justified by 'real-world' risk, there is also – other than in aspects of drug and pesticide application – an almost complete absence of a more utilitarian view in which the benefits of bulk material use are recognised as a contribution to sustainable development. This omission continues to skew the bold vision of the 'Brundtland definition' to which the world signed up in 1987 and further endorsed in 1992, weakly transposing it instead into a dispiriting *de minimus* approach of being a bit less toxic and incrementally more efficient within an otherwise unchallenged but manifestly broken paradigm.

This absence of focus on realisation of benefits was aptly addressed by the 2023 OECD report *Understanding and Applying the Precautionary Principle in the Energy Transition*,[16] commissioned by the European Union specifically concerning application of the PP to address uncertainty surrounding the safety of technologies contributing to reducing carbon emissions and environmental degradation. The OECD report articulates how "*…if applied overly rigidly… the PP can thwart potentially beneficial innovation*". The report notes that conflicts relating to renewable energy technologies, perceived as more apparent and readily quantifiable, tend to blunt focus on their positive contributions to addressing the potentially disastrous effects of unconstrained climate change that are perceived as diffuse, distant in time and less foreseeable. The OECD report notes that "*…regulatory choices that lock in the status quo may not be precautionary at all*", especially when benefits are overlooked and the risks of acting or not acting are balanced. The report notes, in this broader context, that "*…'doing nothing' is often not the most 'precautionary' approach*". There are direct parallels with chemical regulation narrowly fixated on intrinsic properties in isolation, ignoring the wider context of life cycle risk and the ways in which materials, with appropriate cautionary management, can make positive contributions to safely and efficiently meeting human needs.

NOTES

1   ECHA. (n.d.) *Understanding REACH*. European Chemicals Agency, Helsinki. https://echa.europa.eu/regulations/reach/understanding-reach, accessed 28 September 2024.

2 EPA. (n.d.) *Toxic Substances Control Act*. Environmental Protection Agency (EPA). [Online.] https://www.epa.gov/chemicals-under-tsca, accessed 05 October 2024.

3 Sandmann, P.M. (2001). Risk = Hazard + Outrage. *the linkbetween*, 33 (January 2001). [Online.] https://www.psandman.com/articles/zurich.pdf, accessed 25 December 2024.

4 ECHA. (n.d.) *Understanding REACH*. European Chemicals Agency, Helsinki. https://echa.europa.eu/regulations/reach/understanding-reach, accessed 28 September 2024.

5 EcoVadis. (2024). *What is EcoVadis?* EcoVadis. [Online.] https://support. ecovadis.com/hc/en-us/articles/115002531307-What-is-EcoVadis-, accessed 28 September 2024.

6 USGBC. (2024). *LEED Rating System*. US Green Building Council (USGBC). [Online.] https://www.usgbc.org/leed, accessed 28 September 2024.

7 GBI. (2024). *About Green Building Initiative*. Green Building Initiative (GBI). [Online.] https://thegbi.org/, accessed 28 September 2024.

8 BREEAM. (2024). *BREEAM®*. BREEAM. [Online.] https://breeam.com/, accessed 28 September 2024.

9 Green Building Council of Australia. (2024). *What is Green Star?* Green Building Council of Australia. [Online.] https://new.gbca.org.au/green-star/exploring-green-star/, accessed 28 September 2024.

10 Borras, S.M., Franco, J.C., Isakson, S.R., Levidow, L. and Vervest, P. (2015). The rise of flex crops and commodities: implications for research. *The Journal of Peasant Studies*, 43(1), pp. 93–115. https://doi.org/10.1080/03066150.2015. 1036417.

11 van Tulleken, C. (2023). *Ultra-Processed People: Why Do We All Eat Stuff That Isn't Food ... and Why Can't We Stop?* Cornerstone Press, London.

12 Ceddia, M.G. (2020). The super-rich and cropland expansion via direct investments in agriculture. *Nature Sustainability*, 3, pp. 312–318. https://doi.org/10.1038/ s41893-020-0480-2.

13 Gillon, S. (2016). Flexible for whom? Flex crops, crises, fixes and the politics of exchanging use values in US corn production. *The Journal of Peasant Studies*, 43(1), pp. 117–139. https://doi.org/10.1080/03066150.2014.996555.

14 UNDRR. (2025). *What is the Sendai Framework for Disaster Risk Reduction?* United Nations Office for Disaster Risk Reduction (UNDRR). [Online.] https:// www.undrr.org/implementing-sendai-framework/what-sendai-framework, accessed 21 January 2025.

15 Everard, M. and Longhurst, J.W.S. (2018). Reasserting the primacy of human needs to reclaim the 'lost half' of sustainable development. *Science of the Total Environment*, 621, pp. 1243–1254. https://doi.org/10.1016/j.scitotenv.2017.10.104.

16 OECD. (2023). *Understanding and Applying the Precautionary Principle in the Energy Transition*. OECD Publishing, Paris. https://doi.org/10.1787/5b14362c-en.

Innovation for a Very Different Future

6

As Niels Bohr famously said, *"Prediction is very difficult, particularly if it's about the future"*. Every business seeking to orient itself to future markets in a long-term strategy review would surely agree. Tomorrow's business landscape will inevitably be substantially different to today's, and uncertainty is a tough pitch on which business needs to play, innovate and invest.

6.1 THE CHANGING SHAPE OF TOMORROW

We are already encroaching dangerously beyond finite environmental capacities,[1] making the necessity for sustainable innovation unavoidable. A clearer understanding of the biophysical functioning of the world we inhabit, and our multiple interactions with it, can help us better anticipate the nature and likelihood of changes for which innovation will be necessary, be that defensively or for future profitability through how we choose to serve needs.

The increasing depletion of resources will certainly enforce change that goes way beyond simple eco-efficiency. As observed in Chapter 3, humanity's aggregate pressures have become dominant factors bringing about change to biospheric structure, integrity and functioning, driving us into the Anthropocene epoch from the former Holocene during which natural processes comprised dominant influences.[2] Humans have now colonised almost the entire biosphere on this planet Earth and exploited resources at a scale

DOI: 10.1201/9781003637875-6

resulting in flows of material and energy becoming substantially now regulated by socio-economic in addition to the ecological processes that dominated in pre-human times, with progressive evolution through the fundamentally differing socio-metabolic regimes of hunter-gatherers, agrarian societies and into industrial society from which many current global sustainability problems arise.[3] The socio-metabolic qualities of contemporary industrial society are at least as different from a desirable future sustainable society as they are from the earlier agrarian regime. Pursuit of sustainability therefore represents a fundamental re-orientation of societal norms, including the operation of the economy, rather than a matter of simplistic regulatory compliance or marginal technical fixes such as those achievable through narrow promotion of eco-efficiency.

The rising sustainability challenges resulting from resource depletion and disposal will have more prosaic impacts, such as inevitably spirally material costs if founded on depleting resources. Change in markets will also be enforced by societal 'outrage' – what the public may no longer tolerate – as seen for example in the widespread rejection by consumers of many single-use plastic items in the second decade of the twentieth century. Awareness of potential future liabilities emerging from a greater understanding of the environmental and health implications of many substances and use practices may also influence course correction. A change in approach within regulatory regimes is also both necessary and inevitable as society increasingly grasps the threats it now faces. These pressures for change will inevitably accelerate, for example as the grave prognoses of uncontrolled climate change and rapidly diminishing biodiversity are increasingly recognised.

It is better to be forearmed from a backcasting perspective than it is to react as new legal limitations and value chain demands are imposed. Cumulative sustainability pressures will inevitably enforce re-evaluation of traditional resource use habits and corporate priorities, for some strategically and for the laggards reactively, pivoting progressively towards more environmentally and socially responsible and responsive practices.[4] For some sectors of business, elective corporate responsibility policies will shape preparedness for an inevitably much-changed future. We can also expect to see increasing statutory requirements for taking, and reporting on, corporate responsibility to address these inevitably more challenging conditions in the future. Shifts in fiscal policies will also precipitate market reform.

One interesting vision-based approach is the *Ending Plastic Waste* mission developed by CSIRO, Australia's national science agency. Rather than imposing expectations on others, *Ending Plastic Waste* starts with an end goal to which all can agree. From this consensual platform, it then seeks collaborators from across industry, research and government to collectively reimagine societal use of plastics, transform plastic waste into an economic commodity, and "… *create systemic change through data science, materials and manufacturing, recycling processes and whole of life, circular solutions to reduce plastic pollution entering the environment*".[5] Under this mission, CSIRO has set an aspirational target of stopping 85% of leakage of plastics into the environment by 2030, presenting a clear and consensual goal for innovation and collaboration by all sectors of Australian society.

6.2 TRENDS AND SHOCKS

When considering the shifts into future, there are both trends and shocks.

Some trends are already acknowledged and look set to continue. We know, for example, that decarbonisation and energy efficiency will remain with us, as indeed will waste reduction and efforts to halt and ideally reverse serious degradation of biodiversity and its life support processes. Whilst regimes such as the stance of the US in early 2025 aim to abandon such commitments to environmental and social responsibility, biophysical laws will impose their own limitation and adverse impacts including through inevitably increasing climate disruption due to climate-active gases already added to the atmosphere by human activities. Also locked into the system is a trajectory of growth in the human population that is unlikely to stabilise until the middle of the century or possibly later, bringing with it associated demands on resources to meet needs. Further generally accepted trends include demands for greater transparency, increasing elective reporting of sustainability issues, tightening regulatory requirements and more pervasive digital scrutiny.

Then we have 'shocks', comprising unforeseen or poorly understood factors manifesting as disruptions. A history of discovery of adverse toxic effects, climate tipping points, revelation of previously unexplored child labour in supply chains or unanticipated regulatory rules or customer expectations tells us that the emergence of the unpredictable has a habit of being predictable.

In reality, these factors form a continuum, as yesterday's shocks become embedded into the mainstream of societal and business trends.

But, as we look into the hazy crystal ball to explore the future, there are two constants. One is that, as long as humans exist, there will be needs to fulfil. These range from foundational biophysical needs such as food, water, clean air, warmth and shelter and extend to needs for communication, travel, socialisation, freedoms of belief and expression and broader factors that enable people to realise their full potential. The second constant is a fabric of natural laws underpinning the continued functioning of the supporting biosphere with which we co-evolved and upon which all our activities will continue to entirely depend.

There will therefore be continuing demand for products and services, with associated material use, to satisfy society's diverse needs in this changing world. It is therefore essential that, in designing the materials and products necessary to meet these needs, we are fully cognisant of feedback between what we do and the functioning, vitality and finite limits of the wider socio-ecological system within which this occurs. Optimally, we should raise our vision beyond reducing our pressures on ecosystems, working instead to rebuild natural supportive capacities in how we use and manage resources to enhance future security and opportunity, as outlined in the 2020 book *Rebuilding the Earth: Regenerating Our Planet's Life Support Systems for a Sustainable Future.*[6]

Rather than wait for these trends and shocks to disrupt current norms, it makes a great deal more sense to anticipate inevitable constraints on future 'freedoms to operate' driven by the declining carrying capacity of supportive ecosystems impinging upon established commercial habits. To better inform ourselves about this, we can use knowledge about the workings of the biosphere, including limits to its carrying capacity, as a basis for backcasting (backcasting is addressed in greater detail later in this book), such that sustainability principles increasingly inform innovation of new materials and of the ways in which we use substances of all types across societal product life cycles.

6.3 TOOLS FOR THE JOB

As discussed previously, the transposition of the bold intergeneration framing of the 'Brundtland definition' of sustainable development on the meeting of needs both now and into the future has been hampered by a legacy mindset focused principally on reacting to the potential for hazard. Discourse and policy around the meeting of needs are often therefore constrained by concepts such as 'treading lighter on the earth' or being more efficient, based on incrementalism that fails to challenge established norms and approaches. The Brundtland definition's primary focus on meeting needs is largely absent: a 'missing half' of sustainable development and a half that is essential to stimulate innovation and paradigmatic change across all societal activities.

A wide range of tools has been developed and applied globally for the assessment of chemicals by businesses, governments and their regulators. They too have often been limited by historic preoccupation with the potential for hazard and eco-efficiency. This is unfortunate because, as we have seen, considerations about 'real-world' risk across the whole societal life cycles of the products into which materials are incorporated inform us more about their sustainability, including both benefits and disbenefits, than addressing hazards in isolation. This is not to say that these legacy tools and the investments in their application do not serve valuable purposes, as each of them is relevant to specific aspects within the wider picture of sustainable development. However, most are best regarded as 'jigsaw pieces', rather than assumed by extrapolation to represent the whole picture.

Life Cycle Assessment (LCA), for example, is a widely used and valuable approach defined by the international standard ISO 14040 addressing simplified life cycle stages from the sourcing of raw materials right through to the post-use phase (as described in Chapter 4). However, evaluation criteria generally applied under the classic application of LCA tend to focus on negative impacts only – ozone-depleting potential, eutrophication, teratogenic, carcinogenic and mutagenic properties, water and energy inputs and so on – all of which are important considerations but disregard the positive contributions of material use in addressing needs. This omission is significant as addressing needs is, after all, essentially the purpose for which materials are deployed.

A wide range of assessment tools has been developed to support chemical management and regulation. These include the EU REACH approach and supporting Environmental Product Declaration (EPD) and Product Environmental Performance (PEF) metrics, EcoVadis, LEED and Green Star rating systems, and the Green Building Initiative (GBI) and BREEAM® sustainability assessment and accredited standards. Other commonly used approaches to chemical assessment include SciveraLENS®, Greensuite®, GreenScreen List Translator™, GreenWERKS, the Green Chemistry and Commerce Council (GC3) Retailer Database, the OECD Substitution and Alternatives Assessment Toolbox (SAAT), the ECHA Plastic Additives Initiative, Cradle to Cradle (C2C), Additive Sustainability Footprint (ASF), Carbon Handprint, Material Flow Cost Accounting (MFCA) and the GRI 301: Materials approach. In 2022, I published a peer-reviewed scientific paper titled *Assessment of the sustainable use of chemicals on a level playing field*[7] that compared these approaches against a range of principles relevant to the sustainable use of chemicals over whole product life cycles. These principles included: addressing multiple dimensions of sustainability beyond simple chemical characteristics, a foundation in science, consideration of life cycle risk rather than simply intrinsic chemical properties, positive contributions to meeting human needs, open access to tools, whether they are statutory, and that they have been subjected to peer review. This analysis was reproduced and further discussed in my 2024 book *Seeking Sustainable Development on a Level Playing Field: A PVC Case Study.*[8] The analysis of how different chemical assessment approaches map to sustainability-relevant principles was informative and highly germane to the considerations in this book, so the summary table is reproduced here as Table 6.1 as it is relevant to the considerations in this book.

TABLE 6.1   Summary of chemical assessment systems/approaches in terms of their coverage of criteria relevant to sustainable use, emphasised by a 'traffic lights' colour coding (GREEN, Yes, fully meets criterion; AMBER, Partially meets criterion; RED, No, does not meet criterion)

	FULL DIMENSIONS OF SUSTAINABLE DEVELOPMENT	TRANSPARENTLY SCIENCE-BASED	BASED ON FULL ARTICLE LIFE CYCLE RISK (RATHER THAN POTENTIAL HAZARD ALONE)	RECOGNISES POSITIVE CONTRIBUTIONS TO MEETING HUMAN NEEDS	OPEN ACCESS	FREE TO USE (ALBEIT WITH GUIDANCE AND EXTERNAL AUDITING)	APPLICABLE ACROSS PRODUCTS/MATERIALS	STATUTORY	PEER REVIEWED IN SCIENCE LITERATURE
Life Cycle Assessment (LCA)	NO	YES	Partially	NO	Partially	Partially	YES	NO	YES
Environmental Product Declaration (EPD)	NO	YES	Partially	NO	Partially	YES	YES	NO	YES
Product Environmental Footprint (PEF)	NO	YES	Partially	NO	Partially	YES	YES	NO	YES
EU REACH	NO	YES	NO	NO	YES	YES	YES	YES	YES
SciveraLENS®	NO	YES	Partially	NO	NO	NO	YES	NO	NO
Greensuite®	NO	YES	Partially	NO	NO	NO	YES	NO	YES
GreenScreen List Translator™	NO	YES	NO	NO	NO	NO	YES	NO	Partially
GreenWERKS	NO	YES	NO	NO	NO	NO	YES	NO	NO
Green Chemistry and Commerce Council (GC3) Retailer Database	NO	YES	NO	NO	Partially	Partially	YES	NO	NO

(Continued)

TABLE 6.1 (Continued)   Summary of chemical assessment systems/approaches in terms of their coverage of criteria relevant to sustainable use, emphasised by a 'traffic lights' colour coding (GREEN, Yes, fully meets criterion; AMBER, Partially meets criterion; RED, No, does not meet criterion)

	FULL DIMENSIONS OF SUSTAINABLE DEVELOPMENT	TRANSPARENTLY SCIENCE-BASED	BASED ON FULL ARTICLE LIFE CYCLE RISK (RATHER THAN POTENTIAL HAZARD ALONE)	RECOGNISES POSITIVE CONTRIBUTIONS TO MEETING HUMAN NEEDS	OPEN ACCESS	FREE TO USE (ALBEIT WITH GUIDANCE AND EXTERNAL AUDITING)	APPLICABLE ACROSS PRODUCTS/MATERIALS	STATUTORY	PEER REVIEWED IN SCIENCE LITERATURE
OECD Substitution and Alternatives Assessment Toolbox (SAAT)	NO	YES	NO	NO	Partially	Partially	YES	NO	NO
ECHA Plastic Additives Initiative	NO	YES	Partially	NO	YES	YES	Partially	NO	NO
Cradle to Cradle (C2C)	YES	YES	YES	NO	NO	NO	YES	NO	YES
Additive Sustainability Footprint (ASF)	YES	YES	YES	YES	YES	YES	YES	NO	YES
Ecovadis	YES	YES	NO	NO	NO	NO	YES	NO	NO
Carbon Handprint	NO	YES	YES	NO	YES	YES	YES	NO	YES
Material Flow Cost Accounting (MFCA)	NO	YES	Partially	NO	NO	NO	Partially	NO	YES
GRI 301: Materials	NO	YES	NO	NO	**YES**	YES	YES	NO	NO

GREEN, Yes, fully meets criterion; AMBER, Partially meets criterion; RED, No, does not meet criterion.

Most of these approaches have a basis of science, and most are applicable across a range of material types. Few though address multiple dimensions of sustainable development beyond chemical behaviour, and fewer still take a whole product life cycle approach. Only one –ASF – includes an evaluation of the positive benefits provided by the use of substances across the whole product life cycle. ASF was adapted from The Natural Step's (TNS's) Strategic Life Cycle Assessment (SLCA) approach[9] specifically to evaluate and serve as an innovation framework for the sustainable use of additives within whole product life cycles, addressing the four TNS System Conditions covering systematic accumulation of lithospheric and synthetic substances, systemic physical disruption and structural obstacles to people meeting their needs. ASF/SLCA also explicitly explores positive benefits from the use of materials together with negative pressures across that spectrum of issues. The structure of ASF and the wider SLCA approach is generically relevant to the assessment of aspects of the sustainable use of materials across the whole societal value chains of the products within which they are incorporated, consciously founded on exactly the principles articulated above.

This approach is, by intention, generically applicable to the life cycles of materials of all types on a level playing field of sustainability principles. The 2022 paper, with expansion in the subsequent 2024 book, applied the ASF/SLA approach at a generic level to a range of additives in different materials (including metal-based stabiliser additives in PVC compound, brominated flame retardants in polyolefins, preservatives used in timber and cobalt incorporated in some solar panels). Comparative analyses highlighted illustrative examples of pollution-related, resource depletion and ethical issues, addressing both challenges and benefits by TNS System Condition at different stages in product life cycles for each additive type. The conclusions are that all these additives make positive contributions at different life cycle stages (enhancing renewable energy production, promoting product longevity and hence meeting needs more efficiently, etc.) but also generate negative issues requiring innovation (such as ethical issues at the mining stage or inhibition of recycling at product end-of-life).

Extending this principle-based approach further, even 'natural' materials such as paper may have problematic aspects when bleaching during manufacture and the potential for coatings to inhibit value recovery at product end-of-life are encountered. This does rather remind us of the spoiler alert in Chapter 2 that automatic assumptions about inherently sustainable materials are a fantasy when implications across product life are considered.

6.4 DIGITAL FACILITATION FOR SUSTAINABLE TRANSITION

We live in an increasingly digital world that is progressing in leaps and bounds, at least amongst more privileged sectors with access to mobile and internet technologies. Some facets of this interconnected world are driving greater consumerism and dissent, but others are strengthening opportunities for data-driven decision-making as well as greater connection along value chains and inclusive dialogue. Means to track material flows, optimise resource efficiency, learn from innovations and foster transparency across entire value chains has never been greater. This is not the place to unpack all aspects of the accelerating pace and increasing pervasion of the digital revolution, and the opportunities that it presents along with associated risks. However, some instances germane to seeking societal symbiosis for sustainable use of materials are discussed.

Advances in modelling, including visualisation in real time of modelled outputs, can potentially better inform innovators about both the intended and unintended consequences of material choices and enhancements in the context of their use across whole societal value chains, offering immediate feedback to better guide decisions and investments. Suitably trained modelling and visualisation across the whole value chain also enhances opportunities for collaboration between partner businesses as well as legislators, and their consequent cooperation to identify options to improve overall sustainability. This then can feed into communications with customers, NGOs, regulators and wider public interests.

Blockchain technology, comprising growing lists of records (blocks) that are securely linked together via cryptographic hashes with each block containing information about the previous block effectively forming a chain, is most commonly associated with the financial sector and the growing use of cryptocurrencies. However, blockchain is increasingly being implemented in supply chain management. In 2016, technology company IBM launched a platform for companies to enable blockchain record-keeping technology in their supply chains,[10] with Everledger using the technology to track diamonds from mine to store to ensure that what is sold is ethically mined. Further amongst the many ever-proliferating examples of blockchain uses in value chains include tracking oil from well to customer by Abu Dhabi National Oil Company in addition to automating transactions along the value chain, tracking cargo ships and containers by Maersk as well as deliveries by FedEx, tracing tuna from can back to fishermen by John West, and to trace cobalt supplies used in electric car batteries to ensure the authenticity and quality of the product by the Ford Motor Company.[11]

There are many implications for the application of blockchain in the chemical industry and for the wider societal use of chemicals through the linkage of cryptographically coded blocks into chains. Opportunities include verification of the sourcing, use and fate of substances. Blockchain can also inform smart contracts (computer programme or transaction protocols that automatically execute, control or document events and actions), potentially eliminating the need for independent verifiers and building trust and transparency in the handling of materials throughout extended life cycles.[12] Whilst blockchain technology is not without critics and risks, careful design can ensure rigour and security.[13]

Blockchain has direct application to initiatives aiming to improve transparency and accountability in product value chains, including for example underpinning digital product passports (DPPs). From 2024, the European Union implemented a new regulation requiring nearly all products sold in the EU to feature a DPP under the *Ecodesign for Sustainable Products Regulation*, aiming to "*...enhance transparency across product value chains by providing comprehensive information about each product's origin, materials, environmental impact, and disposal recommendations*".[14] DPPs are intended to include a digital record of essential details such as a unique product identifier, regulatory compliance documentation and information on substances of concern, also linked to user manuals, safety instructions and guidance on product disposal. Blockchain can make a significant contribution to linking relevant data transparently across whole product value chains, potentially accelerating the transition towards a circular economy, although issues such as balancing transparency with protecting confidential information need to be resolved. DPPs are likely to become increasingly mainstream in coming years, enhancing the potential for different actors in value chains to work collaboratively to reduce waste and environmental impacts, maximise efficiency and ensure ethical practices that do not infringe land rights, the rights of other people and adverse outcomes for ecosystems.

One of the more prominent digital advances in recent years is the emergence of artificial intelligence (AI), broadly defined as intelligence exhibited by machines and particularly by computer systems. Features of AI include sensitivity to the environment (natural, business or other) and learning to maximise prospects for achieving defined goals. AI is widely used in internet search engines, autonomous vehicles, clinical decision-making, strategy-based games, creative tools (including text, images, music and others) and various forms of decision-support in the face of uncertainty through formulating deductions in the absence of complete knowledge. As with blockchain, this is not the place to explain AI in depth, but rather to illustrate some of the implications and opportunities for the chemical industry and the wider societal use of materials.

Under simpler paradigms when outcomes were more deterministic, the principle of 'putting the rules in the tools' was wise. It is also now more readily applicable to achieve this principle in the face of uncertainty with support from AI. Implications for exploring the consequences of material choice, innovation, handling, use and fate across whole values chains clearly fall into this definition if the AI is trained to explore these ramifications based, for example, on probabilistic evaluation, game theory, reasoning, pattern recognition, for example, through neural networks and sequential learning. Outputs can also be made more intuitive through natural language processing (NLP) interfaces, even responding to questions posed in plain language including through speech recognition. A 2017 survey found that one in five companies reported that they had incorporated AI into some activities, with examples in the fields of energy storage, medical diagnosis, military logistics, predictions of judicial decisions, foreign policy and supply chain management.[15]

A 2024 review observed how AI could create, and already was creating, opportunities ranging from molecule and materials discovery through to new applications and customer acquisition.[16] This potential is significant given the chemical industry's reliance on scientific data for innovation, available albeit imperfect customer data and complicated manufacturing processes. All of these facets can benefit from AI to inform decision-making in the face of uncertainty, accelerate processes and improve efficiency. Generative AI has the potential to accelerate the generation of insights from laboratory data, technical specification sheets, the scientific literature as well as sales presentations, molecular discovery and finding new applications and markets. It can also add insight into the consequences of chemical choices and use along connected value chains. This application of AI has potential knock-on sustainability benefits through the role of the chemical industry in the global economy by providing essential materials for most other industries. All these benefits can be achieved at pace, with lower dependence on the limitations of human interactions beyond the training of the AI and the verification of findings. Balanced against the potential benefits is the environmental footprint of AI, the International Energy Agency forecasting in 2024 that power demand for the use of AI might double by 2026.[17] Whilst AI is being used to make the power grid more efficient and 'intelligent' and to track overall carbon emissions, burgeoning AI data centres already use as much energy as a small country[18] and supercomputing is forecast to increase overall demand for energy that is "...*likely to experience growth not seen in a generation*" consuming a projected 8% of US power by 2030 compared with 3% in 2022.[19]

Whilst rapid evolution in the digital world can be used for good or ill – the proliferation of 'deep fake' and fake news is well known – there is significant potential for the increasing power and pace of modelling, visualisation, blockchain, AI, other breakthroughs and their use in combination to accelerate progress towards sustainability in the chemicals sector as for society as a whole.

6.5 STRATEGIC AND PROFITABLE INNOVATION

Innovation informed by whole life cycle context and sustainability principles, supported by appropriate tools, is essential if we are to deliver the materials and products that satisfy needs sustainably now and into the future. There is, of course, a fine juggling act here, because reaching for perfection today may not be feasible given the limitations of the world we have inherited. Overreaching what the market can bear risks business extinction.

This fine balance between innovation and profitability is one that businesses have to navigate, ideally with support from the policy environment. It is best informed by being clear about strategic direction – the ultimate attainment of sustainability – a vision that can act as a 'pole star' guiding incremental innovations that contribute stepwise in the right direction, albeit that the destination cannot be immediately attained. The wisdom of French Enlightenment writer and philosopher Voltaire (François-Marie Arouet, 1694–1778) – *"Le mieux est l'ennemi du bien"* translating literally to *"The best is the enemy of the good"* – resonates loudly in this context. Reaching too far can blind us to or dismiss the importance of incremental progress. Aggressive NGO campaigning critiquing failure to realise full sustainability in an institutionally unsustainable world can dissuade businesses from investing in or announcing incremental steps heading strategically towards that necessarily distant destination. Inappropriate regulatory demands can also inhibit sustainable progress, as seen when considering conflicts inherent in expectation of the immediate attainment of the two central pillars of European Union's 'Green deal' policy: inflexible implementation of the *Towards a non-toxic environment strategy* that could effectively kill off realisation of the *Circular Economy: Implementation of the Circular Economy Action Plan* condemning society to the dystopian outcome of perpetuating wasteful and polluting linear resource use (see discussion in Chapter 4).

Innovation is a fine balancing act between business, regulators pulling the right levers and promoting the right incentives, societal opinions reflected by and influenced by NGOs and the media, and of course markets that are amoral by nature and need the influence of legislation and other inputs from civil society better to propel sustainable progress.

NOTES

1 Rockström, J., et al. (2009). A safe operating space for humanity. *Nature*, 461(7263), 472–475. https://doi.org/10.1038/461472a.

2 Crutzen, P.J. and Stoermer, E.F. (2000). The 'Anthropocene'. *Global Change Newsletter*, 41, pp. 17–18.

3 Haberl, H., Fischer-Kowalski, M., Krausmann, F., Martinez-Alier, J. and Winiwarter, V. (2011). A socio-metabolic transition towards sustainability? Challenges for another Great Transformation. *Sustainable Development*, 19(1), pp. 1–14. https://doi.org/10.1002/sd.410.

4 UNEP. (2011). *Towards a Green Economy: Pathways to Sustainable Development and Poverty Eradication*. United Nations Environment Programme (UNEP). [Online.] https://www.unep.org/explore-topics/green-economy, accessed 03 October 2024.

5 CSIRO. (2024). *Ending Plastic Waste*. CSIRO. [Online.] https://www.csiro.au/en/about/challenges-missions/ending-plastic-waste, accessed 31 October 2024.

6 Everard, M. (2020). *Rebuilding the Earth: Regenerating Our Planet's Life Support Systems for a Sustainable Future*. Palgrave Macmillan, Cham.

7 Everard, M. (2022). Assessment of the sustainable use of chemicals on a level playing field. *Integrated Environmental Assessment and Management*, 19, pp. 1131–1146. https://doi.org/10.1002/ieam.4723.

8 Everard, M. (2024). *Seeking Sustainable Development on a Level Playing Field: A PVC Case Study*. CRC Press, Boca Raton, FL.

9 Lundholm, K., Blume, R. and Oldmark, J. (2011). *Process Guide to Sustainability Life Cycle Assessment - A strategic approach to assessing the life cycle of product systems using the Framework for Strategic Sustainable Development*. The Natural Step International, Stockholm.

10 Nash, K.S. (2016). IBM pushes blockchain into the supply chain. *The Wall Street Journal*, 14 July 2016. [Online.] https://www.wsj.com/articles/ibm-pushes-blockchain-into-the-supply-chain-1468528824, accessed 06 January 2025.

11 Sristy, A. (2021). Blockchain in the food supply chain - what does the future look like? *Walmart Global Tech*, 30 November 2021. [Online.] https://tech.walmart.com/content/walmart-global-tech/en_us/blog/post/blockchain-in-the-food-supply-chain.html, accessed 06 January 2025.

12 Zhou, X. and Kraft, M. (2022). Blockchain technology in the chemical industry. *Annual Review of Chemical and Biomolecular Engineering*, 13, pp. 347–371. https://doi.org/10.1146/annurev-chembioeng-092120-022935.

13 Bakos, Y., Halaburda, H. and Mueller-Bloch, C. (2021). When permissioned blockchains deliver more decentralization than permissionless. *Communications of the ACM*, 64(2), pp. 20–22. https://doi.org/10.1145/3442371.

14 Digital Product Passport. (2024). *Advancing Transparency and Sustainability*. European Union, 27 September 2024. [Online.] https://data.europa.eu/en/news-events/news/eus-digital-product-passport-advancing-transparency-and-sustainability, accessed 06 January 2025.

15 Ransbotham, S., Kiron, D., Gerbert, P. and Reeves, M. (2017). *Reshaping Business with Artificial Intelligence: Closing the Gap between Ambition and Action.* MIT Sloan Management Review and The Boston Consulting Group, September 2017. [Online.] https://web-assets.bcg.com/img-src/Reshaping%20Business%20 with%20Artificial%20Intelligence_tcm9–177882.pdf, accessed 06 January 2025.

16 Mori, L., Macak, M., Perdur, R.S.M., Wells, I., Gautam, Y., Misra, S. and Green, Z. (2024). *How AI Enables New Possibilities in Chemicals.* McKinsey and Company, 20 November 2024. [Online.] https://www.mckinsey.com/industries/ chemicals/our-insights/how-ai-enables-new-possibilities-in-chemicals, accessed 06 January 2024.

17 IEA. (2024). *Electricity 2024: Analysis and Forecast to 2026.* International Energy Agency (IEA). [Online.] https://www.iea.org/reports/electricity-2024, accessed 06 January 2024.

18 Calvert, B. (2024). AI already uses as much energy as a small country. It's only the beginning. *Vox.com,* 28 March 2024. [Online.] https://www.vox.com/ climate/2024/3/28/24111721/climate-ai-tech-energy-demand-rising, accessed 06 January 2025.

19 Goldman Sachs. (2024). *Generational growth: AI, Data Centers and the Coming US Power Demand Surge.* Goldman Sachs. [Online.] https://www.goldmansachs. com/insights/goldman-sachs-research/generational-growth-ai-data-centers-and-the-coming-us-power-demand-surge, accessed 03 May 2025.

Symbiosis for Sustainability

7

Nature's persistence and constant adaptation and evolution in the face of changing conditions since the genesis of life on Earth over 3.85 billion years ago has occurred through the close co-evolution and integration of all living organisms and processes. Humanity is no different in terms of the rise of civilisations and waves of technical revolutions throughout history, achieved through societal collaboration and differentiation of roles as well as cascading innovation and collective knowledge and capacities. The trace of cultural evolution left by the exploitation or innovation of materials is described in Chapter 2.

There is then a great deal we can learn from the symbiosis of microscopic and macroscopic organisms in the natural world, particularly how the limited genetic complement of multicellular host organisms is augmented by a diversity of microscopic organisms that are not just cohabitants but have evolved in indivisible symbiosis. Microbial constituents adapt rapidly and flexibly to changing diets, environmental conditions, developmental stages and stresses more quickly and efficiently than host organisms can switch genes on and off. In fact, host organisms often entirely lack genes encoding critical pathways necessary for their survival, relying on their symbionts for these essential functions. The same observation about interdependence applies both within and between species, across the kingdoms of plants, animals and fungi, and right up to the interactions upon which biospheric integrity, functioning and stability depend. It also applies in terms of how different but fully interdependent sectors of society interact to either propel or inhibit innovation.

7.1 SUSTAINABLE DEVELOPMENT AND THE SOCIETAL 'SUPERORGANISM'

A pertinent lesson from nature for progress towards the sustainable use of materials is that all four principal societal sectors – private, public, academic, and voluntary – constitute an interactive, potentially and ideally symbiotic, whole

100

DOI: 10.1201/9781003637875-7

entity upon which the stability and progress of civil society depends. If these sectors work antagonistically, the societal ecosystem will be dysfunctional, blocking concerted progress with sustainable development. Like a natural ecosystem, the societal ecosystem needs to function as an integrated system subject as much, or more so, through cooperation rather than competition or antagonism.

Societal cohesion has never been more important than in the present, with sustainable development such an urgent priority for a world with a booming human population depending on much-depleted and still fast-dwindling resources. The interlinked crises of biodiversity, climate and pollution are exacerbated by the dispossession of less powerful people through richer enterprises appropriating resources for financial gain and to supply the demands of more powerful sectors of society. All dimensions of sustainable development require deep collaboration between sectors of society, enterprises within each of these sectors, and recognition that we are all part of the same interactive socio-ecological system and will all ultimately either thrive or suffer together. We are desperately in need of novel symbiotic models within which regulation, business, knowledge generation, activism and consumption across sectors cooperate to make progress towards shared visions of a future that is more secure, profitable and that grants opportunity.

For all our perceived divisions, humanity is an interconnected system whether locally, within river catchments, across land masses and right up to global scale. Complex feedback loops ramify from all of our interactions with supportive ecosystems, for better or for worse, and whether we work divisively or pull together around common goals. In reality, classification of society into different sectors is a purely cultural construct hardly reflecting, for example, how people interact routinely with different elements of this social system and also move between institutions driven not just to seek more pay but taking with them their motivations and life goals. The word 'ecosystem' has at its heart the word 'system', reflecting the breadth and depth of evolved interconnections of all components. So too, the 'socio-ecological system' of which we and all our activities are part, the term describing humanity's complete interdependence with the biosphere with which we co-evolved. And, as we have seen, social metabolism is now a dominant factor in the functioning of the planetary ecosystem.[1] At this parlous point in human history, there may be no more important collective goal than concerted progress towards sustainability. It is therefore important that we address the societal 'superorganism' as a potentially symbiotic whole if we are to accelerate progress towards shared, commonly understood sustainability goals.

7.2 BUSINESS ENGAGEMENT WITH SUSTAINABLE DEVELOPMENT

Although there is a great deal that business institutions can do to address their own environmental and social footprints, there are clear limitations to achieving sustainable material use as this depends upon collaboration across the far wider societal life cycle of products and with policy, fiscal and behavioural environments. Collective action also includes between ostensibly competing businesses, for which collaboration already occurs in terms of pooling resources for research and innovation and for amplifying influence on markets, supply chains and the policy environment. Collaboration with disparate partners in other sectors connected with product value chains is also essential if materials are to be used sustainably. This includes not just other business segments but also the policy and market environments within which value chains operate and that either facilitate or inhibit sustainable use. Progress towards sustainability ultimately requires substantial symbiotic activity across all societal sectors.

As the capitalist system has been either chosen or accepted by much of global society as the dominant trading model for turning raw materials into useful products and services addressing human needs,[2] it is therefore the business sector that we primarily depend upon for innovations in material types, uses and recovery to meet our needs in a future that will inevitably be increasingly constrained by the conflict of growing human demands and declining natural resources and assimilative capacities. If we are to take seriously the essence of the 'Brundtland definition' of sustainable development, we need the creative energies of business to innovate ways to meet the material needs of current and future generations on a safe, efficient and profitable basis.

Business is absolutely necessary for innovation and is subject to an analogue to the laws of natural selection as it profits or dies under imperfect market forces. Consequently, innovation is in the DNA of leading players in the corporate sector, and it is to business that we look for the novel materials, products and use patterns required to serve needs in the face of today's daunting sustainability challenges. One thing I have certainly learned in dealing with a wide range of forward-thinking businesses over many years is that, contrary to some more critical perspectives, it is far from a simple case of regulatory and NGO sectors being 'on the side of the angels' necessary to constrain otherwise predatory corporate activities! Rather, we absolutely depend on the corporate sector as a primary source of innovation and investment in new approaches, constituting vital underpinnings on the journey towards sustainability.

7.2.1 Initial Externalisation of Environmental and Social Consequences

The contemporary resource use model and the economic and market systems that underpin it are largely a product of the protracted period of time now referred to as the European Industrial Revolution. Even today, these anachronistic systems remain substantially unreconstructed. The global human population of approximately half-a-billion people at the commencement of the European Industrial Revolution was only around one-sixteenth of contemporary numbers, and the world's capacity to supply natural resources and absorb wastes then also seemed boundless. The funding of empire-building, principally by European nations, saw colonies established, a primary driving factor of which was resource acquisition making up for depletion or shortfalls at home. The principal limiting factor to economic progress in this new industrial era was a shortage of people, both as consumers to purchase products from newly founded industries and also to serve as labour. Extensive mechanisation of agriculture freed up labour from the land, enabling mass migration to burgeoning industrial cities to bolster the workforce required to serve growing manufacturing enterprises.

This was also a time when industry was working with relatively simple chemistry, certainly as compared to the complex materials in widespread use in society today. In this far-gone time, environmental limits such as those we see today threatening climate stability and the collapse of biodiversity were far less pressing, and the perspective of a bountiful and boundless Earth did not imbue a culture of resource conservation, equity and what we now term as sustainability.

The narrow financial capture of business was exacerbated through progressive neoliberalisation. This capture was substantially accelerated with global pervasion by the influence of US economist Milton Friedman, a politically influential advocate of free-market capitalism unconstrained by state intervention under a programme collectively known as 'monetarism'. Friedman won the 1976 Nobel Memorial Prize for Economic Sciences for his work, which significantly influenced US and UK policy with global ramifications in globalising markets throughout and since the 1980s, reshaping modern capitalism. However, the vision that the role of business was purely to make money, 'liberated' from environmental, commerce, social and other constraints, is dangerously narrow. It has been a direct contributor to 'offshoring' of manufacturing and other activities to regions with less rigorous environmental and ethical standards that nevertheless constitute significant elements of the overall sustainability footprint of consumer goods. It has also widened disparities between rich asset owners and the poor, both within and between nations. Overall, it has accelerated overexploitation and serial degradation of primary environmental and social resources, insulating profit-taking from liabilities stemming from the means deployed for generation of profit. Fundamentally, a narrow monetarist worldview – shaping 1980s culture in which 'greed is good' became a mantra – overlooks the fact that markets are fundamentally amoral. It thereby ignores the fact that protective restraints are essential not only to avert damaging collateral impacts but also to give progressive businesses the confidence to invest in socially and environmentally responsible innovation consistent with broad global consensus and rhetoric about commitment to a sustainable pathway of development.

7.2.2 Corporate Reengagement with Sustainable Development

Within these narrow financial blinkers, backed up by statutory duties on the corporate sector to maximise returns to shareholders, subsequent re-engagement of businesses with what we now know as the environmental and social dimensions of sustainable development was often one of incomprehension and denial. Regulation of business activities was often perceived as a constraint on enterprise, as environmental and social issues were almost entirely externalised under the monetarist dogma.

Sweden is often rightly recognised as pioneering a proactive approach to the uptake of sustainable development. In his 2002 book The *Natural Step Story: Seeding a Quiet Revolution*,[3] Professor Karl-Henrik Robèrt, the founder of The Natural Step, credits me with analysing the reasons why this would be so. Robèrt cites one of my articles in the *Stepping Stones* newsletter that I used to write, in which I noted that Sweden's industry had until relatively recently been overwhelmingly forest-based and so relied on long-term planting. But, around the beginning of the twentieth century, Sweden essentially ran out of trees and was also co-incidentally hit by a devastating famine that claimed many lives. Swedish society consequently had faced in its recent cultural history the consequences of a lack of foresight in balancing its needs with the supportive capacities of the environment. This awareness built upon the already prevalent need for greater social cohesion to deal with the hostile climate visited upon Scandinavia for much of the year. Whatever the causative factors, Sweden certainly did begin to engage proactively with sustainable development earlier and more comprehensively than many other developed-world countries.

7.2.3 The Golden Age of Altruism

Business, though, has grasped aspects of its social and environmental responsibilities from far further back in history. Narratives about the European
Industrial Revolution often reference low-wage employment, hazardous technologies and practices, child labour and formerly unprecedented levels of pollution. However, running counter to these images is another reality that this,
at least in Industrialising England, was also an age of unprecedented altruism. This 'golden age of altruism' saw museums, schools, healthcare facilities,
parks, libraries and zoos founded as philanthropic ventures funded by captains
of industry enjoying formerly unprecedented levels of wealth. These altruistic
ventures constitute an early form of payback to society, recirculating newly
found wealth towards the wellbeing of urban communities driving the wheels
of novel and profitable industry.

Philanthropy has continued to be associated with recirculation of wealth
by some richer members of society, generally directed by their private interests. Many social and environmental contributions have been thus funded.
Investment by the Packard family, formerly of Packard Bell fame, promoted
marine science and conservation by underwriting the founding of the Monterey
Bay Aquarium in Cannery Row (Monterey, California) as well as the adjacent
Monterey Bay Aquarium Research Institute.[4,5] Famously, the Bill and Melinda
Gates Foundation continues to invest heavily into environmental and medical schemes around the world intended to contribute to positive outcomes for
humanity.[6]

This form of philanthropy is inherently a private venture by individuals
and consortia of wealthier members of society, albeit now often moderated
by quasi-independent boards, returning wealth to selected humanitarian and
environmental causes. Arguably, corporate donations to environmental and
human development non-governmental organisations (NGOs) are another
modern reflection of this pattern of philanthropic giving, albeit that some are
justly accused of constituting 'greenwash' or 'hush money' if not well-directed
and transparently audited.

7.2.4 Supply Chain Security

Although the age of empire was one of resource grabs from a seemingly
boundless planet, issues of supply chain security have also shaped business
awareness of the need for stewardship of some finite supplies of resources.
Growing societal awareness and concern, NGO and media exposés of issues

such as child and indentured labour often distantly down supply chains, as well as degradation or destruction of vulnerable habitats contributing to declines in species, have subsequently exerted pressure on businesses to explore their supply chains. For some more foresighted companies, supply chain stewardship became an issue driven by corporate responsibility. For laggards, regulation, market and shareholder pressures or public outrage became principal drivers of change.

Various businesses have been key partners in the instigation of independently audited stewardship schemes accrediting environmentally and socially responsible sourcing from raw material acquisition through to final markets in material and product life cycles. This is particularly so for large corporations that have been in existence for a matter of a century or more, often with more deeply rooted cultures of long-term planning. Some leading examples of supply chain stewardship schemes instigated with business involvement are summarised in Box 7.1. These and other stewardship schemes have created market differentiation and platforms that are accessible to all players in that supply chain, influencing consumer choice and behaviour and reshaping markets and societal norms.

BOX 7.1:   EXAMPLES OF SUPPLY CHAIN STEWARDSHIP SCHEMES INITIATED WITH BUSINESS INVOLVEMENT

- The Forest Stewardship Council (FSC) brand was instigated by a cross-societal consortium of founding organisations, including the Kingfisher Group (owners of the B&Q chain of do-it-yourself stores in the UK) as well as the NGO Greenpeace and a disparate group of other players including representatives of forest-based communities. FSC was established in 1994 as an international, non-governmental organisation setting auditable standards for supply chains for wood, paper and other forest products. It has since grown into the world's most widespread forest certification system covering more than 160 million hectares of forest.[7]
- The Marine Stewardship Council (MSC) was instigated by a consortium of founding partners including the multinational corporation Unilever (the world's largest company purchasing fish to serve domestic markets) to promote sustainable exploitation of fishery stocks securing stock for current and future generations. This multi-party venture became fully operational in 1999, established with a vision of "...the world's

oceans teeming with life, and seafood supplies safeguarded for this and future generations". MSC is based on similar principles to the FSC, including an ecolabel and fishery certification program. In 2024, over 15 million tonnes, or 19%, of all wild marine catch across 36 countries globally had engaged with the MSC programme.[8]

- The Aquaculture Stewardship Council (ASC) is seeking to emulate the MSC and FSC approaches for the 60% of seafood eaten around the world that is farmed.[9] The ASC was founded in 2010 by the World Wide Fund for Nature (WWF) and the Dutch Sustainable Trade Initiative (IDH) with the aim of managing and implementing socially responsible aquaculture.
- The Rainforest Alliance brand was founded in 1987 with the mission of creating a world where people and nature thrive in harmony, instituting a certification scheme wherein brands can advertise that published environmental and ethical standards have been observed throughout supply chains.[10]
- A range of traceability and supply chain performance standards are found in the food sector. The UK-based Red Tractor certification scheme is one of many initiatives around the world assuring that certain standards are met in food supply chains. Red Tractor standards relate to the traceability of food and how it is farmed, transported and packed and are verified by dedicated inspectors.[11]
- At least some companies are being approached to supply Vegan (animal-free) chemical substances including biofuels, feeding back down supply chains to meet the demands of customers and consumers.
- The Roundtable on Sustainable Palm Oil (RSPO)[12] and the Round Table on Responsible Soy (RTRS),[13] addressed in Chapter 4, are not without their critics but have been established by diverse stakeholders to establish environmental and social standards and codes of conduct applied across entire supply chains.
- Supply chain assurance has extended to wider initiatives such as the Conflict Minerals Reporting Template developed by Responsible Minerals Initiative, against which organisations can assess and transparently declare their ethical responsibility to their downstream customers with respect to 3TG (tin, tantalum, tungsten and gold).[14]

- A growing number of companies purchasing cocoa have made public commitments to address sustainability issues in their value chains, generally acting through trading companies. However, research using shipping data from eight countries responsible for 80% of global cocoa exports has revealed that the overall adoption of public sustainability commitments is low, with just over one quarter (26%) of cocoa traded under some form of sustainability commitment. Traceability and transparency are necessary features of robust and effective voluntary sustainability commitments, particularly with respect to extending current efforts to smaller traders and indirect suppliers.[15]
- The multinational home products company Unilever announced commitments in 2024 to undertake revisions of its supply chain to "...*help build a more equitable and inclusive society by raising living standards across its value chain*", creating opportunities through inclusivity, and preparing people for the future of work by ensuring all providers of goods and services earn at least a living wage or income by 2030, and a commitment to spend €2 billion annually with suppliers owned and managed by people from under-represented groups by 2025.[16]
- The Better Cotton Initiative, described in more detail in Chapter 4, seeks to engage whole cotton value chains to promote more sustainable cotton supply and use.[17]

7.2.5 Changing Paradigms of Producer Responsibility

There was a time when the perceived responsibilities of chemical substance producers were limited to production facilities and direct emissions alone. The regulatory environment in advanced economies has, over time, resulted in enhanced management of production processes, though a narrow focus on production facilities alone still persists in some geographies and companies. Progressively, business concerns about wider environmental issues such as water use and aqueous pollutants began to be influenced by public acceptability and, eventually, regulatory controls. Social issues such as noise, odour and transport disruption have progressively been factored into regulatory and elective requirements. Awareness of responsibilities 'downstream' of production sites has also seeped into corporate consciousness and practice, whether to avert liabilities or as a proactive approach to responsible business.

Initially, responsibilities perceived by many chemical manufacturing companies tended to end with Safety Data Sheets (SDSs) in Europe or similar instruments elsewhere, specifying properties and appropriate use of manufactured substances. Responsibility though often ended there, with the consequences of the use of substances delegated down the value chain.

Extended responsibility began to take shape under concepts such as 'Product Stewardship' and 'Extended Producer Responsibility', identifying issues further downstream in the value chain and seeking longer-term solutions to the management of waste products. This took a more formal shape in the chemicals sector with the launch of the Responsible Care® programme in 1985 by the Chemistry Industry Association of Canada (formerly the Canadian Chemical Producers' Association). The Responsible Care programme has subsequently grown into a voluntary initiative covering 67 countries with combined chemical industries accounting for nearly 90% of global chemical production.[18] Health, safety and environmental performance are the primary focal areas of the Responsible Care programme, under which signatory chemical companies agree to continuous improvement of performance. Though not a legal requirement – and critics have argued that a lack of sanctions and reliance on self-regulation has led to some behaviours in contradiction of the programme's stated intent[19] – voluntary commitments under the Responsible Care Programme are complementary to government regulation.

A range of foresighted businesses have also invested in product and material collection, sorting and recycling schemes to promote increasingly sustainable resource use when products reach the end of their useful service lives. Examples of progressive 'downstream' schemes and initiatives promoted by businesses are summarised in Box 7.2, demonstrating increasing ownership of corporate connections in value chains across the wider socio-ecological system within which these enterprises operate and upon which they rely for primary resources and continuing societal 'permission to operate'.

BOX 7.2:   EXAMPLES OF DOWNSTREAM VALUE CHAIN INVESTMENTS BY BUSINESSES PROMOTING CIRCULARITY

- High recycling rates for various high-value metals (gold, copper and silver) as well as glass and paper have been addressed in Chapter 4. These recycling rates have been achieved by players in the value chain collaborating and pooling resources to put in place relevant infrastructure available to all businesses in the value chains of these materials.

- The polyvinyl chloride (PVC) value chain also has a significant level of corporate collaboration to fund recovery and recycling infrastructure. In Europe, many initiatives relating to the promotion of cyclic use of end-of-life medical devices, flooring, window profiles and wiring are led under the VinylPlus® voluntary commitment.[20] Similar cyclic PVC schemes are also established in other territories, for example the PVC Stewardship Scheme (PSP) in Australia.[21]

- Printer ink cartridge recycling is well-established, internet searches revealing a diversity of free or rewarding take-back and recycling schemes.

- Batteries and electrical waste are also well catered-for in terms of take-back and recovery schemes in various territories across the world. Value recovery is a major driver as is the diversion of potentially hazardous substances from mixed municipal waste flows, for example in the UK.[22]

- There is growing activity around the recycling of post-consumer cotton. Worn clothes with a high cotton percentage disposed of at end-of-life are sorted for reuse and recycled after sorting by colour and removal of non-fabric elements then blended with virgin cotton to compensate for their reduced fibre length or alternatively they are downcycled.[23] These practices reduce landfill or other forms of waste disposal, recovering value and reducing dependence on virgin resources.

7.2.6 The Rise of Corporate Social Responsibility

A logical progression from private philanthropy and producer responsibility is a more structured approach at corporate level. Corporate social responsibility (CSR) has emerged as one approach better-linking business activities with their social and environmental roots, variously as voluntary or statutory requirements in different parts of the world.

CSR is widely seen as contributing to brand reputation[24] and the perception of doing business in a socially responsible way.[25] Ideally, CSR activities should align closely with corporate strategy and impacts, enabling the corporate sector to identify and take ownership of its wider footprints. If not, it tends to be regarded as tokenism or, at worse, 'greenwash'.[26] Optimally, the aim is not just to offset harm, but to take a regenerative approach to primary environmental and social dependencies, as framed by the position of the outdoor clothing company Patagonia, Inc. that harming the environment is ethically objectionable.[27]

In 2019, nearly 200 chief executives in the US formed a Roundtable declaring their fundamental commitment to serving all stakeholders and to 'do good', including not just for shareholders but also for employees, suppliers, customers and communities, constituting a radical commitment to CSR.[28] In 2020, 96% of the world's largest 250 companies and 80% of large firms worldwide reported on their sustainability performance.[29] Mandatory CSR reporting can have significant capital market benefits as well as altering a business' social and environmental impact, though it is not without risks as success is conditional on well-designed reporting rules.[30] Lack of standards for sustainability reporting in the US is an obstacle to the value of CSR reporting and the degree to which is it trusted. The Global Reporting Initiative[31] and the International Financial Reporting Standards Foundation[32] are two approaches seeking to address this gap through the development of consistent standards.

CSR is one of many initiatives seeking to drive a more sustainable accommodation of profit generation with environmental and social interdependencies in the corporate sector. Perception and practice of CSR have evolved over time from one of self-regulation internal to businesses towards external declaration of pledges to go beyond legal requirements aimed at delivering beneficial impacts on local communities and the environment. Internal reporting on environmental and social implications of business still continues, generally referred to as environmental, social and governance (ESG), as a guide to direct corporate investments in a more responsible direction.[33]

7.2.7 Voluntary Commitments and Vision-Based Planning

A successful model that has emerged in terms of capturing the innovative potential of business is the pursuit of clearly stated voluntary commitments. Voluntary codes of conduct comprise commitments voluntarily made by organisations or consortia of enterprises seeking common goals without legal compulsions. They can make these commitments for the benefit of themselves, collaboratively across business sectors, as well as for wider communities including consumers, workers and citizens. A key aspect of voluntary commitments is that they take the form of visions that a company or business sector intends to achieve, serving as a navigation point set in the future against which to guide interim plans that represent steps towards the achievement of these visions.

Voluntary commitments are robust if transparently founded on credible and firmly founded sustainability frameworks – for example, the Sustainable Development Goals (SDGs), The Natural Step framework, Planetary Boundaries or 'Doughnut Economics' – and it is also necessary for them to have associated relevant and measurable targets. This then provides a framework against which industry can innovate. It may also enable regulators to agree that the goals are consistent with societal aspirations, to check that promises are being kept and that audited progress towards clearly stated targets is being made, allowing space within this strategic framing for novel approaches.

One exciting project I was involved in was at the planning stage of the Great Western Hospital in Swindon (Wiltshire, England), for which a sustainability vision was developed and a sustainability action plan was subsequently created. Subcontractors bidding for different work packages entailed in the hospital build were explicitly required to include a costed element relating to how they would contribute to this sustainability action plan. Training was offered to prospective bidders about the sustainability model (The Natural Step approach) upon which the vision and action plan were based; bids were rejected if they failed to include sustainable innovation. The net result was substantial innovation around various aspects of the build, including, for example: a novel drywall system that has subsequently become an industry standard; enhanced roof insulation reducing the need for some radiators and decreasing energy consumption over the hospital life cycle; and a different approach to painting that reduced the number of coats required as well as improving wash water efficiency. It would be fatuous to claim that the Great Western Hospital is fully sustainable, but progress towards the goal of sustainability – sustainable development – was certainly achieved through presenting a robustly founded vision against which contractors were required to innovate.

The United Nations Global Compact published a *Growing into Your Sustainability Commitments: A Roadmap for Impact and Value Creation* report in 2013 profiling how companies integrate voluntary sustainability commitments into their strategies and operations, serving as a framework to help companies navigate their engagement with and derive value from voluntary sustainability commitments.[34]

7.2.8 Redefining the Purpose of Business

Ultimately, this comes down to corporate purpose. If a business defines itself merely by maximising return on investment to shareholders – a common mantra of narrowly framed business models founded on Milton Friedman's monetarist model of corporate governance underpinning much of the Thatcherite and Reaganomics models of business from the 1980s – then issues such as environmental responsibility and social care are seen simply as inhibitors to aggressive corporate and international competition. This is very much manifest in the campaigning and subsequent strategies imposed in 2025 by US President Donald Trump, with the *Make America Great Again* slogan (which remains conveniently unclear about when it was previously great for all sectors of American society) narrowly framed on economic competition, particularly with China, to be achieved by dissolving social and environmental standards and international agreements, and limiting the redistribution of benefits across society as these are regarded as unnecessary and limiting 'red tape' on the maximisation of profit. The myopia of monetarism is alive and well, also evident in modern India, Brazil, China, Turkey and various other major economies.

As we have seen throughout history, the capitalist business model dominant today in the industrialised world is one that is being accepted, actively or passively, as the primary means by which society converts raw materials into useful products to serve its diverse needs. As we head into a more contested future, limited by environmental capacity to support the demands of growing human numbers, sustainability pressures will inevitably impinge ever more tightly on our freedom to operate. Far from representing unnecessary 'red tape', social and environmental concerns and limitations will frame future markets and routes to profitability. It is the businesses that target the meeting of needs as an outcome that will accrue more stable returns on investments in an inevitably changing operating environment, profitably delivering against those needs in the most efficient and safe manner.

Innovation in society comes from multiple sources. Traditionally, we might have thought of government leadership setting a direction of travel or research organisations determining issues of emerging concern. However, whilst all in society have roles as change agents towards a sustainable future, it is the business sector that puts material flesh on conceptual bones and is therefore a principal change agent in the redirection of society. A diverse community of enterprises is lumped together under the monolithic term 'business', spanning players in societal value chains from the raw material extraction phase, through material manufacturing and right the way through to waste disposal and recycling operators as well as advertising agencies, media, retailers, wholesalers and consultants. Without the agency of all these business enterprises and their associated investments, the potential to join up along the value chain is not going to happen. Businesses, therefore, need to revise their vision, acknowledging the importance of collaboration for mutual benefit along value chains that can prosper through being more united under collective visions about the achievement of sustainable use of specific materials or products.

The rise of 'green consumerism', particularly prominent in the 1980s, has built a legacy of branded products found today on supermarket shelves (such as Rainforest Alliance, FSC and MSC marques, discussed elsewhere in this chapter). This legacy indicates that green consumerism, at least for those who can differentiate and afford it, is alive and well. The advertising sector has a significant role to play in deploying its skill in stimulating desire and consequently demand, exerting a significant influence on consumption habits across society. It is tiresome to see frequent advertisements on the television telling us that we should aspire to a new mobile telephone every year, when the ones we may have had for several years serve their function perfectly adequately: this is a negative form of advertising from a sustainability point of view, driving hyper-consumption and inevitably leaving in its wake mounds of electronic waste with all of its embedded materials, energy and value. The advertising sector and those that commission it have responsibilities here in messaging to the wider public that needs can be met more efficiently and safely and that durable products have a significant role to play not only in conferring long-term value upon us individually as well as collectively through reducing our overall environmental and social footprints. This might be linked to what is best for the world, although altruism is not necessarily a dominant feature of all markets. However, bulk trade markets, for example, in selecting materials from which products in the built environment are made, are amenable to key messages about long-term durability, low maintenance requirements, reduction in associated costs and reduced potential liabilities through a greater focus on responsible sourcing through to low-hazard and high potential for value recovery and recycling. Stimulus of demand will in turn inspire innovation of increasingly sustainable materials and products, progressively redirecting the mainstream of societal consumption habits.

Failure to observe the objective reality of the sustainability challenges facing the world, as a global issue but also bearing down on procurement decisions and personal life satisfaction, is a failure to serve the needs of society now and into the future. This would be in direct contravention to often-repeated rhetorical commitments to sustainable development by businesses and government alike. This shortfall needs to be recognised, and ideally also regulated, to support openly stated commitments in a sustainable pathway of development. Failure of government to date to impose such regulations is an indicator of the power of vested business interests in influencing policy to perpetuate open profiteering from unsustainable norms.

Across all societal sectors, marketing has the power to build common shared visions about a sustainable future, offering security and promise to all. This is a challenge that the marketing sector should grasp responsibly as a key propellant towards a sustainable future better serving our needs now and tomorrow. This transition is essential if we are to pivot away from inherited market norms founded substantially on driving excessive and unnecessary consumerism, particularly of 'false satisfiers'.

7.3 REGULATORY ENGAGEMENT WITH SUSTAINABLE DEVELOPMENT

A brief overview of aspects of the progressive evolution of environmental and social regulation pertaining to chemical production and use, germane also to wider human activities, is featured in Chapter 2. A key aspect of the history and much of the current 'state of play' of regulation is a predominant focus on negative factors: essentially of 'being less bad'. Setting and enforcing statutory minimum performance has been, and remains, important to control less progressive or recalcitrant members of society. However, 'bottom-up' regulation alone is an insufficient model to drive innovation towards better meeting needs in a more constrained future.

7.3.1 From Piecemeal Control to Strategic Vision

Reaction to emerging issues of concern on a piecemeal basis fails to convey a necessary sense of cohesion around addressing future challenges. When faced with a diversity of strands of regulatory requirements that lack cohesion, staff within businesses – small businesses in particular – perceive an apparent jumble of obligations and wider concerns. These range from energy reduction and decarbonisation to primary resource depletion, potential human rights violations in supply chains including identifying slave, child or indentured labour, as well as emissions to all environment media in manufacturing. Further requirements also pertain to addressing community concerns, issuing comprehensive SDSs for all chemical products, wider engagement with producer responsibility, and auditing of water and energy use. There is also a catalogue of legislation relating to substances of concern as well as workers' rights, design for cyclic recovery, and many more. Without a strategic vision, this array of issues may appear bafflingly disparate and be perceived as net constraints on wealth-generating activities.

The consequence of lack of clarity about how these facets relate together, and why they matter, was evident when I was in the regulatory sector in the early 1990s. The language of BATNEEC had recently entered the lexicon of the environmental regulation of businesses. BATNEEC is an acronym for 'Best Available Technology Not Entailing Excessive Cost', a laudable attempt to balance safeguarding the environment with the economic feasibility of implementing appropriate pollution control measures, sharing burdens between the economic interests of businesses and the interests of other sectors of society. Interpretation and application were inevitably context-dependent. BATNEEC was more generally referred to conversationally as CATNIP – 'Cheapest Available Technology Not Involving Prosecution' – reflecting the reluctance of businesses to go beyond *de minimus* action to meet regulation!

A piecemeal approach also tends to lead to 'regrettable substitutions' as organisations look to evade today's problems without foresight. Substitution of substances identified as 'bad' by regulation or by media or NGO campaigning does not automatically mean that alternative substances are automatically 'good' or 'better', highlighting a fundamental flaw in the assumption that sustainable development can be driven solely by attempts to sequentially reduce harm over time without challenging the underlying paradigm of material use throughout whole life cycles. Yet a great deal of planning, by business and in government, has historically been mired in evading today's problems without broader consideration of the systemic ramifications of perceived solutions, falling into the trap of investing in tomorrow's as-yet untested problems. Our current heavy dependence on chemical inputs in agriculture is a case in point. As we have seen there has been a carousel of substitution of pesticide substances from arsenic to methyl bromide, organophosphates, organochlorine substances, synthetic pyrethroids and neonicotinoids to name a few broad categories. Each new 'wonder chemical' arrived with great promise but only subsequently were problems to emerge, leading to a new generation of 'magic bullets' that also promised great things but in turn were also found to be problematic for similar or other reasons, and so the carousel turned again to new untested solutions. If we frame the problem as avoiding damage from current chemical use – and let us not forget the power of agribusiness in maintaining profit by selling new products that perpetuate this approach – we almost inevitably forecast our way into investment in tomorrow's as-yet undiscovered problems. Framing the problem instead as one of meeting the need for crop protection in the safest and most efficient way might stimulate broader thinking about novel approaches, such as the use of solarisation for soil sterilisation, introduction of pest predators or reinstating traditional rotational practices that break parasite and other pest life cycles. Based on emerging learning, we might also consider stimulation of the microbiome of plants and animals as a solution to averting or reducing the need for pesticides or veterinary drugs, achieved, for example, through the introduction of prebiotics (compounds that foster growth or activity of beneficial microorganisms) or seeding with probiotics (live microorganisms conferring benefits to host organisms).

Other examples in this book illustrate where a narrowly framed forecasting approach founded on extrapolation of today's norms has led to new problems through regrettable substitutions. A further classic example is that of the replacement of chlorofluorocarbons (CFCs) with hydrochlorofluorocarbons (HCFCs) as refrigerants and propellants after the ozone-depleting properties of CFCs became apparent. Although representing an easy step for the chemical industry, and also compatible with pre-existing refrigerant technologies, HCFC replacements were also ozone-depleting (albeit to a lesser extent than CFCs) but a greater contributor to climate change. As another example, I was invited to speak with the Board of a major global sportswear brand and was told at the outset of the meeting that the company had decided to phase out a particular substance that was attracting media attention. Clearly, the Board expected me to be impressed, but first I asked what they had replaced the substance with. This question was met with incomprehension. Clearly, this kneejerk substitution approach could merely lead this company to invest instead in tomorrow's untested potential problem; a meaningful conversation ensued! This experience was also a frightening revelation of how senior captains of a leading global business brand could be so myopic, with an associated dangerously narrow perception of corporate risk. This is not to say that establishing baselines of pollution control and ethical performance are unimportant – they absolutely are – but that this approach in isolation does not propel thinking, innovation and investment towards a clearly articulated vision of a sustainable future that may be radically different from today's norms, informing what this may entail for society's material use and the markets that are required to serve it. Clearly, an issue-by-issue approach limits the perception and promotion of the bold vision of sustainable development about the meeting of needs of both current and future generations as set out in 1987, and signed up to globally, in the Brundtland definition. Novel, vision-led approaches are necessary to promote innovation to meet the inevitably different needs of people in a fast-changing future.

7.3.2 Stimulation of Innovation through Strategic Signals

Whilst the statutory sector may not be the primary innovator of materials and products serving changing societal needs, it nonetheless can play an important role in setting directions for travel. When considering voluntary commitments and audited targets in the business sector, it was noted that regulators could develop a partnership mode of operating based on agreement about strategic goals, checking that linked voluntary promises are being kept and that audited progress towards clearly stated targets is being made. Notwithstanding the need to retain a punitive 'bottom-up' compliance regime when dealing with laggards, this is a different regulatory model framed around an agreed vision and aspirations consistent with the Brundtland emphasis on striving to meet needs now and tomorrow. It is a paradigmatic change towards a more symbiotic relationship allowing space for innovation and confidence for investment by regulated businesses.

Some 'green shoots' of a vision-led approach are beginning to appear in strategic government documents at national as well as intergovernmental levels, offering businesses greater confidence to invest in novel approaches. A current example of direction-setting is the stated intent of various national governments to phase out petrochemically powered vehicles, although this carries with it vestiges of the 'bottom-up' paradigm by its focus on phasing out a particular technology, also assuming that further innovation in the internal combustion engine cannot represent a stepping stone towards alternative fuels. A more logical strategic vision against which to backcast and innovate would have been to focus on the achievement of zero, or as close to zero as possible, climate-active emissions rather than focusing on one particular technology. Wider strategic statements about achieving 'Net Zero' across society do though now constitute a logical vision against which to frame innovation pathways, liberating entrepreneurs to imagine and innovate novel technologies and to provide them with the confidence to invest. A signal-led approach is certainly a more efficient stimulant of innovation than the traditional punitive approach of 'bottom-up' regulation. This is because not only the punitive approach is enacted only when harm has already happened, and vested interests have already been established and most likely will be defended, but also it fails to challenge dominant material use and business paradigms. A proactive and enabling policy environment is essential to nurture voluntary approaches from businesses, whether individually or through consortia sharing common interests.

Further enabling strategic signals are emerging from the statutory sector, setting 'directions of travel'. In Europe, for example, regulations such as the 1997 *End-of-Life Vehicles Directive* and its subsequent amendments and the 2012 *Waste Electrical and Electronic Equipment Directive* (the 'WEEE Directive') provide future targets against which innovation and investment can be directed towards more inherently recyclable materials and products as well as in take-back and recycling infrastructure and markets for recyclate. The EU's 2018 *Circular Economy: Implementation of the Circular Economy Action Plan* is one of several national and regional visions seeking to increase the circularity of materials used in society, also providing direction for innovations and investments in industry that are necessary for practical delivery of the goal. Further examples of regulatory instruments offering strategic signals include the European Commission's *Ecodesign for Sustainable Products Regulation*[35] and, at the national level, stringent waste prevention and management regulation under the French 'Circular Economy Law' aimed at promoting a circular economic model based on the eco-design of products, responsible consumption, extension of shelf-life, reuse of products and the recycling of waste.[36]

It is important to understand what 'strategy' means. The term has many dictionary definitions but, essentially, it relates to establishing goals and priorities from which progressive actions can be planned for their eventual attainment. An over-zealous approach expecting the immediate fulfilment of strategic goals – when the old paradigm of regulatory expectation overrides the liberating intent of a longer-term strategic approach – can stifle rather than stimulate progressive action. Conflicts between the EU's 'clean chemistry' and 'circular economy' aspirations in the absence of understanding that a longer-term glide path is essential for their eventual resolution are discussed in Chapter 4.

Ideally, a strategic regulatory approach would set out sustainability aspirations in broad but non-prescriptive terms, dovetailing with robustly founded voluntary sustainable development strategies by businesses and business sectors with regular monitoring of both compliance with basic requirements but also progress against stated goals. Visionary approaches also need to be matched with shifting regulatory scrutineering and ideally subsidies or other incentives. This would be an example of business/regulator symbiosis in action. The benefits of greater regulator-business sector symbiosis are clear, though deliberately codesigned examples of such partnerships are few today.

7.3.3 The Evolution of CSR Linking Voluntary Commitments to Regulatory Signals

There is a need to further develop the relationship between a vision-led approach in leading businesses engaged in voluntary commitments with emergent strategic signalling by regulators at national, regional or intergovernmental levels. One example is found in the way that CSR has evolved from an elective business-led approach into one that is now being formalised and structured by mandatory requirements, albeit in a variable way across the world.

Under Section 135 of India's Companies Act, 2013, for example, qualifying corporations are required to make mandatory financial contributions directed to projects or programmes contributing to a better society and a clean environment. The required contribution is at least 2% of net profits, calculated as an average of the three immediately preceding financial years. Targeting of CSR is under recommendation by a CSR committee and submitted for approval by the company's Board of Directors.

In Europe, the Corporate Social Responsibility Directive (CSRD) requires large companies (>500 employees) and other companies listed on stock exchanges to publish regular reports on their social and environmental risks and impact, with the aim of helping investors, civil society organisations, consumers and other stakeholders to evaluate the sustainability performance of companies. The CSRD entered force in 2023 and there will be phased implementation of large companies and subsequently of SMEs (small and medium-sized enterprises) listed on EU-regulated markets. Essentially, CSRD is an update of the 2014 EU's *Non-Financial Reporting Directive* (NFRD), which applied only to certain large companies and had relatively loose reporting requirements. CSRD spans more companies, requires assurance of information through auditing and is based on a structured mandatory European Sustainability Reporting Standard (ESRS). CSRD is intended to bring reporting of sustainability-relevant information up to a parallel level with established financial reporting, which covers issues such as profit, loss and financial risks. Its intended beneficiaries are the investment community to better understand sustainability impacts and risks, to institutions in civil society by making accountability visible, and for reporting companies by helping them develop more resilient strategies taking account of sustainability implications.

The Australian Government mandated sustainability reporting under the *Treasury Laws Amendment (Financial Market Infrastructure and Other Measures) Bill*,[37] which entered force in January 2025. The Act mandates sustainability reporting to be included in annual reports. Initially restricted only to climate risks, the law will expand in future to cover other sustainability topics such as nature and biodiversity when relevant standards have been developed.

Further legislation such as the European Union's Deforestation Directive (*Regulation on Deforestation-free Products*[38]) is intended to "*...guarantee that the products EU citizens consume do not contribute to deforestation or forest degradation worldwide*". This will drive businesses not only to audit their supply chains with regard to the physical destruction of nature and the displacement of human rights but also to innovate biodiversity-neutral or regenerative practices.

7.3.4 Backcasting as a Strategic Approach

Fundamental to strategic planning, contrasting with common forecasting-based reaction to what is perceived as 'bad' today without adequate foresight, is to ask what 'good' actually means. Backcasting, already mentioned in various places in this book, takes its reference point where we aspire to be in future, and not where we are today. Another name that has been applied to this approach is 'future history creation'. The goal may not be immediately attainable or indeed achievable even in the medium term, but it sets a strategic goal against which contributory innovations can be plotted in a stepwise manner. This goal then acts as guidance regarding what constitutes systemic progress towards sustainability. Stimulus of new thinking founded on the achievement of desirable and consensual long-term outcomes is exactly what is required more generally for progression towards a sustainable future. From visions of what the end goal entails, we can plot actions that are attainable in the short and medium term that lead progressively towards the desired destination, addressing obstacles between where we are today and where we need to be tomorrow.

This approach is also used in politics as, for example, in China's famed long-term planning. The *Global China 2049 Initiative* aims to *"…build a modern socialist country that is prosperous, strong, democratic, culturally advanced and harmonious"*.[39] It is also used in psychology in the form of visualisation (also known as 'mental imagery' in psychological literature) entailing conjuring images in the mind of a desirable goal or state. These images are vividly conceived by engaging all the senses in projecting what that future state feels like, some aspects recalled from memories but others imagined, leading to stimulation of ideas and pathways leading towards its attainment. Visualisation techniques are widely used in sports as well as some aspects of business and personal development. They also have applications for treating mental health through 'neural overlap' or, in other words, wilfully planning for a clearly visualised desirable future outcome rather than being trapped in the legacy of past experience. Notwithstanding some scepticism about these techniques, research based on behavioural, brain imaging and clinical assessments supports the claim that depictive internal visual imagery functions like a weak form of perception with a potentially significant role in mental processes with potential for novel therapeutic approaches.[40] In self-development and pseudo-science approaches, 'manifesting' has been a popular approach, based on the notion that picturing something in the mind (visualising) and backing that up with affirmation (for example repeating positive phrases) can facilitate thinking dreams into reality.

Whether in long-term planning in China or elsewhere in business, or within the reframing of futures in the heads of individuals for therapeutic or self-development goals, this generic approach is founded on backcasting. Key features are that a clear image of a desirable future is fleshed out with as much vivid detail as possible, and then steps leading progressively in the direction of its achievement are also conceived. Backcasting is a valuable means to support companies in breaking from entrenched habits and norms by creating a goal-informed space for innovation of novel thinking, products and processes. From a sustainable development perspective, it is of course essential that visions are founded robustly on solid science-based principles grounding what sustainability means. Vision development and backcasting are more powerful when companies connected along value chains work together around shared, mutually beneficial aspirations.

Even more significantly, co-development of shared visions of outcomes can also link societal sectors together, as seen above where strategic regulatory signals can be matched by corporate voluntary commitments. This cross-sectoral sharing of vision is vital if corporate innovation to better meet needs profitably is to be matched by enabling legislative, fiscal and ideally market environments. It can also inspire consumers regarding the benefits of greater sustainability in the practical and relatable terms of needs met more safely, with fewer problematic issues down the value chain, greater durability and service life conferring optimisation of benefits, and averting legacy problems and costs when products reach the end of their useful lives.

7.3.5 Backcasting for Sustainability

One of the clear distinctions between reacting to today's problems and embarking on the journey of sustainable development is that, contrary to the happenchance of our industrial development to date, there has to be a degree of intentionality and a clear vision articulation of the goal – sustainability – to which we are headed. The goal of achieving a state of sustainability is distant from the unsustainable norms of the contemporary world, but it is possible to develop visions founded on robust and consensual scientific sustainability principles. Sustainable visions can be structured on a range of systemic frameworks including, for example, The Natural Step, the SDGs, Doughnut Economics, Planetary Boundaries and STEEP (social, technological, environmental, economic, political). Whatever the conceptual framework used, clarity about often distant but scientifically founded outcomes is an essential feature of the backcasting approach. As observed with consideration of visualisation, the more vivid and multi-sensory the vision invoked, and the more clearly stepwise progress towards it is conceived, the greater the likely traction.

The World Business Council for Sustainable Development advocates that *"A complete life cycle approach that accounts for environmental and social impacts can help to deliver the cohesive systems transformations required to address the climate emergency, nature loss and mounting inequality"*, advancing a 'Products & Materials' pathway as one of nine transformational pathways at the heart of its *Vision 2050: Time to Transform*.[41] The Products & Materials pathway relates to *"'Things' – the goods people use to fulfil their needs and aspirations, and the assets and materials businesses need to operate and grow"*, noting that *"Profound systems transformation will not be brought about by sticking to our existing ideas and priorities"* as it outlines the need to move from risk mitigation towards a fully regenerative approach.[42] This need for a mindset change to break with the former *de minimus* paradigm of incrementally being less harmful, refocusing instead on a more distant horizon of regeneration of the primary environmental and social resources underpinning future business security, is consistent with the work of other sustainable development proponents and influencers such as themes addressed in my 2020 book *Rebuilding the Earth: Regenerating Our Planet's Life Support Systems for a Sustainable Future*.[43]

At a wider societal scale, Bob Costanza advocates taking a 'sustainable wellbeing future' approach in his 2023 book *Addicted to Growth*.[44] This approach is founded on a robust vision that all in society can co-create and buy into as a more attractive aspiration than current norms, mired as they are by inherent 'addiction' to consumerism and inherent unsustainability. Although far from perfect, Costanza suggests that the 17 UN SDGs and their 169 associated targets are the closest consensual definition that the world has yet been able to agree upon as a set of desirable outcomes for sustainable development and that they therefore serve as an accepted 'dashboard' for monitoring the likely contributions or hazards consequent from policies, innovations and strategies. This chimes with the need to be guided by meeting the needs of the future consistent with the 'Brundtland definition' articulation of sustainable development, as opposed to the narrow legacy framing of reducing harm within an otherwise unreconstructed paradigm. It is, though, important that vivid visualisations relevant to all sectors of society are conceived such that the visions are relatable to the daily experiences, values and priorities of all who engage in their co-creation.

Downscaling bold intentions concerning an increasingly sustainable pathway of development founded on future needs has largely yet to be achieved when it comes to material use. Yet, in all domains of chemical choice, innovation and use, a backcasting approach is essential, as the means by which human and environmental needs are fulfilled is central to the diverse but tightly interlinked challenges entailed in navigating towards a sustainable future. Some pioneering examples of sustainability-related strategies backcasting from clear end goals are emerging are outlined in Box 7.3.

BOX 7.3:   EXAMPLES OF SUSTAINABILITY-RELATED STRATEGIES BACKCASTING FROM CLEAR END GOALS

- Public outcry in the late 2010s, reinforced by the 'Attenborough effect' of media images of wildlife in trouble in the face of gross accumulation of marine litter, created near-global outrage and calls to phase out single-use plastic items in marine litter. Intent to move towards the end goal of halting the accumulation of persistent marine litter drove changes in consumer choice and expectations, implementation of legislation and fiscal measures, as well as proactivity on the part of many businesses voluntarily phasing out single-use plastic applications.

- In Germany, the Energiewende ('energy transformation') programme was instituted with fiscal and other measures as a clear-sighted longer-term commitment to decarbonise the economy, phase out nuclear generation and add political security.[45] Though the programme is not without its critics, it contributed to converting almost 30% of Germany's electrical generating capacity to solar and wind power by 2015 from virtually a zero baseline and also drove down costs to help these renewable generation technologies achieve mainstream market penetration.[46] By the first quarter of 2024, renewable sources met 56% of Germany's power consumption indicating continuing progress.[47] Russia's invasion of Ukraine in 2022 forced Germany, as well as Italy, to revise energy policies further increasing support for renewable energy and passing necessary regulatory reforms.[48]

- 'Net zero' is widely articulated in global, national and regional rhetoric with respect to attenuating climate-active emissions, representing a clear end-point for backcasting for which a range of measures (emission reduction, carbon sequestration in soils and habitats, carbon capture and storage and many more options) can be implemented flexibly to build towards the stated goal.

- The *Ending Plastic Waste* mission developed by CSIRO (Australia's national science agency) also takes a future-oriented approach towards the clear goal articulated in its title. Led by this research body, collaborators are sought from across multiple sectors of society – industry, research and government – recognising that substantial co-development is required to reimagine and exert tangible influence on the societal use of plastics, transform plastic waste into an economic commodity, and "... *create systemic change through data science, materials and manufacturing, recycling processes and whole of life, circular solutions to reduce plastic pollution entering the environment*".[49]

- The European Union 'Green deal' policy and its two central pillars (*Towards a non-toxic environment strategy* and the *Circular Economy: Implementation of the Circular Economy Action Plan*) also represent conceptual end-points, notwithstanding discussion in Chapter 4 about their inevitable internal conflicts if longer-term aspirations are converted into unrealistic immediate expectations.

- Key players along the PVC value chain in Europe have combined interests and funding streams under the VinylPlus® voluntary commitment to sustainable development.[50] The VinylPlus voluntary commitments to 2020 were adapted from Five TNS Challenges for PVC developed from backcasting using The Natural Step tools published in 2000.[51] The Five TNS Challenges for PVC are now subsumed within three VinylPlus 'Pathways to 2030', but still remain explicit in targets for which progress is reported and audited. Backcasting from a vision of sustainable use of PVC, the programme has driven major innovation and sustainable progress including measures such as the voluntary phase-out of cadmium and lead in stabilisers, increasing volumes of recycling of post-consumer waste, elimination of fugitive emissions and increasing energy and carbon efficiency, amongst others.

- Examples of sustainable visioning of other material value chains integrated from raw material extraction through to post-use are elusive, though the paper industry has both upstream (sustainable sourcing and product stewardship schemes to product market) and downstream (promotion of recycling infrastructure and markets) communities and delivery mechanisms albeit that these are not conceived as a fully integrated whole.

- Internet searches around terms such as 'sustainable use of materials' return many initiatives relating to waste handling and mining of novel resources, but lack a full life cycle perspective. There is also a focus on Sustainable Consumption and Production (known as SCP), defined by UNEP as "*...about doing more and better with less. It is also about decoupling economic growth from environmental degradation, increasing resource efficiency and promoting sustainable lifestyles*",[52] though without presenting a clear and scientifically founded vision against which to backcast. Other articulations include the Sustainable Materials Management

(SMM) approach as advocated for example by the US EPA,[53] recognising the need to decouple economic growth from material consumption, but aspirationally as a basis for continuous improvement though without an explicit end-point against which to aim.

- Examples are published of the circular use of polyolefins (including, for example, polyethylene as blown film converted into bags that can be washed after use before recycling to pellets[54]), though single-use or short-life applications tend to result in contamination and difficulties for recovery and sorting. Furthermore, cross-linked polymers of this and other types cannot be simply mechanically recycled (e.g. melted and re-extruded of calendered), requiring instead intensive chemical and energy inputs compromising their potential for genuinely circular use.
- Gold, as a high-value example of metal cycling, has both upstream supply chain stewardship (see discussion of 3TG in Chapter 4) and a significant economic value resulting in a high degree of recovery and recycling, and these generally disconnected initiatives nonetheless address the whole life cycle.
- The worlds of fashion and the clothing industry are associated with substantial polluting and ethical issues in extended international supply chains as well as waste generation particularly through 'fast fashion'. In this sector though, increasing attention is now being paid, particularly amongst 'designer labels', to reduce footprints not only in supply chains but also in addressing end-of-life reuse. This includes, for example, new business models such as clothes rental in which consumers purchase the service of the clothes but not ownership of the physical product. 'Zero waste' clothing brands using circular practices are emerging, using both upcycled and recycled materials and also ensuring that upcycled garments are in turn beneficially recovered.[55] The Sustainable Jungle Brand Rating & Review System comprises a 22-point rating system spanning four key criteria (Nature & Animals, Communities & Wellbeing, Business Values & Governance and Product Performance) informed by publicly accessible data (including disclosure on corporate websites, audits, certifications and public comments) as well as direct reviews,

generating an 'overall rating' as a percentage score as well as benchmarking with other brands.[56] Sustainable Jungle lists companies including Hernest Project, Malaika New York, Whimsey + Row, Anekdot, MUD Jeans, Re/Done and Beyond Retro that are engaged with this programme. Whilst there is not a clear end-point against which to backcast, and influence on the apparel sector is mainly via the market rather than cross-corporate or cross-sectoral, performance and improvement goals set a direction of travel for both upstream and downstream performance.

- In addition to corporate activities, there is growing interest in the regulatory sector on requirements regarding designing for recycling that will become increasingly prevalent. As one example, the Circular Plastics Alliance is an initiative under the European Strategy for Plastics (in particular, under Annex III related to voluntary pledges by industry) to help plastic value chains boost the EU market for recycled plastics to 10 million tonnes by 2025, with over 200 organisations from full plastics value chains signed up by 2020.[57]

- The *Global Commitment: a common vision for a circular economy for plastics* programme, launched in 2018 by the Ellen Macarthur Foundation and the United Nations Environment Programme (UNEO), seeks to "...*unite businesses, governments, NGOs, and investors behind a common vision of a circular economy in which we eliminate the plastic we don't need; innovate towards new materials and business models; and circulate all the plastic we still use, to keep it in the economy and out of the environment*".[58] This is a cross-sectoral, goal-oriented approach to work towards a clear end goal of halting plastic pollution, setting a vision backed by 1,000 organisations at the time of writing. The Ellen MacArthur Foundation and the WWF are convening a *Business Coalition for a Global Plastics Treaty* aimed at developing an ambitious, effective and legally binding UN treaty to end plastic pollution in collaboration with aligned businesses and supported by strategic NGO partners.[59]

- Sale of services rather than ownership of physical products has been explored by a number of business sectors, creating an incentive for the circular reuse of materials recovered when products reach end-of-life. A service-based approach to leasing carpet tiles and other floor coverings has been pioneered by companies such as the multinational company interface, for example through its EverGreen Lease™ scheme that eliminates initial capital outlay and embraces responsible environmental stewardship as worm carpets tiles are recovered for reuse or remanufacture using the company's ReEntry recycling program.[60]

- The European Union's Water Framework Directive is a piece of legislation that sets out rules to halt deterioration in the status of EU water bodies and also an aspiration to achieve Good Ecological Status (GES) for Europe's rivers, lakes, groundwater, estuaries and coastal waters. Although GES is defined by a range of chemical, biological and hydromorphic parameters, the means for its achievement is not specified allowing a flexible approach under which European Member States identify periodic cycles of Programmes of Measures leading towards the attainment of this goal. This therefore represents a form of backcasting that is not presumptive about the means of meeting locally relevant goals.

- Sustainability Victoria, a statutory agency of the Australian state of Victoria established under the Sustainability Victoria Act 2005 (SV Act), has a remit to shape Victoria's circular economy on behalf of the state government. It is intended that this should be achieved through partnerships across industry and the community, with the intent of reducing waste or stopping it at source in every part of the system. A significant contribution to this goal is SV's *Recycled First Policy*[61] requiring all tenderers on Victorian major transport projects to demonstrate within their bids "...*how they will optimise the use of recycled and reused materials at the levels allowed under current standards and specifications*", also inviting them to identify opportunities to "...*trial new innovative products or opportunities to boost recycled and reused material quantities within existing standards and specifications*". Using the weight of public procurement helps achieve the scale necessary to make the handling of waste and the creation of beneficial recyclate economically attractive, not only contributing to more sustainable transport infrastructure outcomes but also generating offerings for wider non-state markets.

It is right that regulators do not lose focus on constant improvement of the bottom line of environmental and social performance, with punitive action taken against those that pollute or exploit. However, there is a need to break away from a divisive perception of businesses as villains and regulators as heroes, and of overzealous or over-important regulatory attitudes and narrowly framed NGO campaigning as well as overly defensive behaviours in business. Antagonism between business and policy/regulatory sectors merely imposes barriers to progress towards commonly conceived sustainable outcomes. At a level of strategy, it is vital that these two private and public pillars are mutually supportive if society is to accelerate sustainable development initiatives to a pace anywhere near commensurate with the gravity of threat from evitable changes in an uncertain future.

Further evolution of regulatory approaches is essential, progressing beyond but still maintaining control of a socially acceptable bottom line of performance. The 'missing half' of understanding and transposition of the concept of sustainable development – a regulatory approach that enables and rewards innovation to better meet needs – requires serious development in the policy environment, including vision-led legislation and associated fiscal measures, if it is to promote sustainable innovation to better address societal goals. A strategic, outcome-based reframing of regulatory approaches would provide direction and certainty against which businesses can innovate and invest.

This reformed model of an open and facilitating approach contrasts with the currently slow process of drafting statutory bottom-up regulations, which are necessarily bogged down by political machinations and inevitably entail trade-offs. Furthermore, the need for a solid evidence base to establish accepted statutory minima generally means that regulations enter statute only retrospectively, perhaps even on a decadal basis. This is witnessed, for example, in the tortuous pathway towards imposition of statutory controls on the horticultural trade and distribution of invasive plant species only after they have been proven to cause substantial disruption when the genie is effectively loosed from the bottle and the plants are firmly established generally beyond effective control. The same is true of chemicals for which deeply entrenched vested interests may have been established at commercialisation, therefore requiring major political will to challenge. We see this, for example, in the 'grandfathering' of substances already in use before the US EPA was charged under the Toxic Substances Control Act to regulate largely novel substances.

It is logical that the private business and public regulatory strands should meet to stimulate innovation. Synergies between enabling strategic signals and linked innovation in business committed through voluntary but auditable approaches are necessary to stimulate innovation and associated investment by the private sector, for example in recovery and recycling infrastructure that is essential to progress towards a circular economy. Collaboration between sectors, rather than antagonism, is a necessary stimulus of innovation for sustainable development, although it will still be necessary to undertake traditional bottom-up regulatory approaches to address the *de minimus* performance of laggards and less responsible businesses. Stringent standards will also be required to manage imports from producers in regions that may compete financially through lower regard for environmental and social ramifications than those accepted by local producers.

There will probably always be disagreements on issues of detail between innovators and regulators. However, a vision-based approach is a more powerful stimulant of innovation and practical progress towards consensual aspirations than the anachronistic perspective of business as villain and regulation as hero, arising as an unhelpful and false dichotomy paralleling that of uncritically assumed 'good' or 'bad' materials. Rather, the synergy between businesses and regulation in co-creative partnership can unite efforts to innovate for sustainability around agreed long-term visions. Greater cross-sectoral consensus is vital if we are to succeed in embracing the 'missing half' of sustainable development relating to inspiring innovation to better address the needs of a sustainable future.

7.3.6 Redefining the Purpose of Regulation

As with business, questions of regulatory purpose must be considered in the light of what constitutes the promotion of stated commitment to sustainable development, taking account of the 'missing half' of stimulating innovation towards consensual goals as well as enforcing 'bottom-up' standards to cater for less progressive businesses. Does the regulatory model contribute to progressing novel, safe and efficient ways of meeting needs in an inevitably more resource-constrained future? Alternatively, if unreconstructed, does it stifle paradigmatic change and the acceleration of progress at this challenging point in history? Undoubtedly, a new symbiotic regulatory model is needed that transparently works across societal sectors to support progress towards consensual sustainability goals.

A further necessity is that regulators bear down on advertising and unfounded environmental claims. The European Union recognised that the lack of specific rules on the claimed 'green' nature of products can mislead or confuse consumers. To address this gap, the European Commission advanced a proposal in March 2023 for a Directive on green claims that would require companies to "*...substantiate the voluntary green claims they make in business-to-consumer commercial practices, by complying with a number of requirements regarding their assessment*" including explicitly "*taking a* *life-cycle perspective*".[62] The Directive has not progressed into statute at the time of writing, but it is an example of legislators and regulators around the world working to add rigour to pros-sustainability claims. Adding rigour is necessary to provide confidence bolstering the continuing permeation of 'green consumerism', emerging particularly from the 1980s, that has built a legacy of branded products found today on supermarket shelves (such as Rainforest Alliance, FSC and MSC marques, discussed elsewhere in this book) as well as promoting scrutiny and verification along supporting value chains. Advertisers making these claims as well as direct communications from producing businesses should be subject to increasing regulation.

7.4 CROSS-SECTORAL SYMBIOSIS FOR SUSTAINABLE DEVELOPMENT

Planet Earth is strange. Its atmosphere is not only dense but also extraordinarily unstable. Why on earth (literally) does it comprise 20% of the highly reactive element oxygen, when simple thermodynamic theory suggests it should not? Carbon dioxide and other constituents also act as a 'greenhouse' maintaining a warmer temperature than would otherwise theoretically occur at this orbital distance from the sun, but not so hot that water cannot exist in liquid form. This awareness of extraordinary instability was noted by atmospheric chemist James Lovelock as a signature of the agency of life on this 'third rock from the sun', comprising a signal to look for in his work searching for extraterrestrial life. But it is Lovelock's extension of this work into the Gaia theory, proposed in the 1960s and developed with the US biologist Lynn Margulis in the 1970s, for which he is best known. Essentially, under the Gaia theory, it is life that moderates the planetary environment, cumulatively acting homeostatically to create conditions amenable to the continued existence and evolution of life. Water, nutrients, minerals, energy and organic matter all cycle through webs comprising predators and prey, parasites and hosts, shredders and decomposers, and plants thriving in the soil that countless microorganisms and other small organisms build. At the planetary scale, life is a coherent 'superorganism' comprising many competing, collaborative and complementary elements working in a deeply interconnected way to maintain the integrity of the whole.

Humanity is also essentially a complex 'superorganism' comprising a diversity of synergistically interacting individuals of the same species maintaining a greater whole, be that at anything from institutional or local scales with their distinct cultures up to global levels, all levels of which fall under a broad definition of 'society'. Social structures existed in prehistory. However, they developed substantially after people overcame the primary limiting factor of daily food sufficiency through collaborative agriculture, water management and trading, which liberated energies and creativity to differentiate social roles many of which had not formerly existed. Exploitation of new and novel materials further accelerated cultural progression and the differentiation and establishment of social roles. There are, of course, significant differences in the drivers and interactions within social systems when compared with the deep and far from fully understood complexities of ecological systems. Amongst these societal drivers is that public perception of 'truth' seems increasingly less objective in the twenty-first-century post-truth era. Perhaps, it was ever thus in former times when superstitions and religious or political diktat prevailed more strongly, but public opinions are now shaped by appeals to emotion and personal belief beyond the boundaries of facts. Adding to this heady mix of pressures and diversions driving societal decisions and actions is the power of vested interests that undermines perceptions of pressing challenges, such as climate change or biodiversity collapse, portraying them as constraints to 'progress' under a flawed model of unfettered neoliberalism. However, humanity is just one component of broader socio-ecological systems, again at scales from the local and regional up to the global, and societies progress or fall with the integrity of supportive ecosystems as well as through internal tensions within society. Jared Diamond's masterful 2011 book *Collapse: How Societies Choose to Fail or Survive* outlines, through many examples, how environmental damage, generally resulting from conflict between short-term interests and long-term sustainability, climate change, hostile neighbours, loss of trading partners and society's response to these challenges can lead to some societies successfully adapting to environmental pressures whilst others collapse due to a combination of factors often exacerbated by cultural and political choices or inertia.[63]

Human societies are of course complex entities. But, just as the simplification of resource life cycle thinking provides a useful basis for tackling complexity in that context (see Chapter 4), so too a simplified classification of societal structure can help navigate the complexities of societal interactions. Some classifications of society are based on economic sectors, social class and combined socio-economic metrics (such as the National Statistics Socio-Economic Classification used in the UK[64]). More usefully, a meta-level classification of society into four broad categories – private (business), public governance, knowledge creation and civil society organisations – serves as a simplification of a host of religious, architectural, communicating, legislative, educational, healthcare, nurturing, trading and other processes upon which the stable functioning and progression of the whole of society depends. There is of course complexity within each of these categories (for example 'civil society organisations' span a range of formal and less formal organisations giving voices to different interest groups including that of minorities and are sometimes referred to as the 'fourth sector') as well as interactions between them. However, in common with the ostensibly disparate competitive, collaborative and complementary strands permeating natural systems, these diverse societal facets – resource exploiters, manufacturers, regulators and rule-makers, protestors, champions of environmental and social interests, farmers, consumers, intellectuals, consultants and many more – combine at a meta level into a coherent whole. Whether at the scale of the planetary biosphere or the machinations of humanity, there is, despite the apparent conflict and complexity, a coherent and ideally cohesive entirety. Nature's durability and progression over 3.85 billion years shows us that collaboration for sustainability is achievable. For humanity today, a significant degree of intentionality and urgency is required to reorient ourselves coherently to attain this same goal.

A handshake between innovators in business and regulators is one of the key connections that need to evolve if we are to harness energies across society to accelerate progress towards sustainability. We have also to consider and integrate other linkages between the four principal divisions of business, regulation, civil society organisations and knowledge-providers such as those in academia and more strategic consultancies.

7.4.1 From Antagonism to Synergy

Historically, and in the present, there is a tendency for these sectors to work antagonistically. The example of anachronistic bottom-up regulation, as already discussed, essentially engenders a culture in which regulators treat business with distrust rather than stimulating and supporting necessary innovation to ensure that people's needs are met safely and efficiently in an inevitably more resource-constricted future. The same is true of the voluntary sector, with many early environmental and social campaigning NGOs railing against damaging corporate activities as well as failure to legislate or take action against offenders. This upwelling of civic unrest certainly served a valuable and necessary role in less enlightened times, constituting the launchpad of the 'environment movement', and it has since led onwards to growing corporate and government awareness of, and response to and adoption of, sustainability issues. A confrontational approach from regulators and NGOs is still necessary to challenge laggards that seek to maximise profit through irresponsible behaviours, and NGO campaigning can also be influential in challenging regulations that inhibit progress towards sustainability. The knowledge-providing sector has formerly tended to study issues within disciplinary boundaries, rendering the transfer of knowledge into the inherently messy real world difficult, as well as often failing to address externalities for other disciplines. These frictions and fragmentation of effort may have promoted initial awareness of challenges with wider societal awareness of, and engagement with, sustainability challenges risks stifling progressive and strategic action if we just expend our energies on what we disagree about in the here and now. We all, though, wherever we are situated in this four-cornered classification, use products and services generated by business so we are all complicit in both problems and potential solutions relating to the use of materials.

As noted above, there will probably always be disagreements on issues of detail between innovators and regulators, and the same is true from the standpoints of the voluntary sector and the knowledge-providing community. However, just as a shared, vision-based approach is recognised as a more powerful stimulant of innovation and practical progress between innovators and regulators, it is also the case that greater collaboration can lead to the energies and insights of campaigning or solution-oriented NGOs and researchers being more strategically directed towards visions developed from longer-term consensual goals. The need for intentionality-directed progress towards a clear vision of the goal of sustainability is widely recognised. It must now be translated into how all in society can converge around agreed desirable outcomes against which to invest their energies and investments. Rather than infighting in the present, what are the goals that we can identify and the visions we can co-create around them to work more symbiotically towards a sustainable future? Without a systemically informed vision, we are likely only to perpetuate our current unsustainable trajectory that ultimately will only exhaust and extinguish itself. Better then to recognise that we need to harness the unique qualities of all societal sectors towards accelerating progress towards common goals.

So, how do we go about reorienting and accelerating progress towards the sustainable use of materials? The promise of 'Backcasting for sustainability', and the risks of failing so to do, have been articulated previously. If we are not to waste the potential benefits bestowed upon us by our shared but dwindling resource base, including through the beneficial reuse of resources already deployed by society, we need to be informed by a common, scientifically grounded vision of what exactly constitutes the sustainable use of materials. Throughout the pages of this book, we have elaborated some key principles necessary to build this vision:

1. A systemic approach to sustainability principles is essential, including addressing chemical, physical and wider socio-economic ramifications of decisions and actions as an interdependent whole;
2. Considering material use within the context of the whole societal life cycles of the products into which they are integrated to frame 'real world' risks and potential benefits;

3. A precautionary approach that addresses meeting the needs of today and tomorrow as strategic goals, averting the risk of losing sight of this bigger picture through narrow preoccupation with immediate potential for hazard;

4. Acknowledging that sustainable development is a journey requiring symbiotic collaboration between all societal sectors, as decisions and investments in wise material use include innovation in business, proactivity in regulatory approach, shifts in fiscal and market levers, investments in civil infrastructure, smart campaigning and communication and systemic insights; and also

5. Expectations recognising that we start from today's highly unsustainable norms, such that shortfalls in achieving immediate perfect fulfilment of goals do not blind us to the importance of making and rewarding stepwise progress in the right strategic direction.

Each of these elements is challenging; achieving them simultaneously is more so. However, this is the nature of culture change. Meaningful shift in material use across society is a key vector of this societal transition onto a pathway of development shaped by sustainability principles. Running throughout the principles outlined above is the need for greater synergy both within and between societal sectors.

It is not incremental agreement over fine details in the immediate term that will enable us to collaborate most effectively, but shared recognition about common aspirations in the longer term; in other words, shared visions that can serve as 'pole stars' to guide and integrate energies and investments across society. Ultimately, it is a collective aspiration to achieve a sustainable future that binds all sectors of society, rising above contemporary factional differences and friction over divisive issues of detail in the contemporary unsustainable world.

7.4.2 Building Collective Vision

The global aspiration of sustainable development is already deeply embedded in rhetoric from governments, voluntary organisations and businesses at all scales. However, there remains a gap in terms of the practical downscaling of this grand ideal into day-to-day decisions and experiences, expressing them in aspirational and inspirational ways that are graphic and meaningful not just for the world as a whole but are also relatable in practical terms for all sectors and individuals. Adding flesh and colour at these pragmatic levels is necessary to guide understanding and engagement by people from all sectors of society. This is necessary to engender acceptance as well as preferential procurement of products made from materials that are more responsibly sourced, durable requiring low maintenance inputs, and then recoverable at end-of-life without potential liabilities and costs associated with waste generation or threats to environmental or human health.

Relevant learning in this regard comes from my work integrating nature conservation, water security and rural development in India and parts of Africa. In international development interventions, I am always starkly aware that I am a white foreigner in other people's land. Imposition of a diktat that people 'should' conserve species and other natural resources plays badly as neocolonial arrogance. This patronising approach justifiably gains no traction, not merely as we in the industrialised world are no exemplars, having undertaken a destructive development journey, but because people's livelihoods and aspirations are intimately integrated with the geographical, ecological and cultural settings they inhabit. The progressive resolution is to work with local people to articulate the practical meaning of nature conservation in terms of how it benefits those who are key actors in its stewardship. Practical examples include the ramifications of protecting ecosystems and their functioning for, amongst other outcomes, greater water and food water security underpinning self-sufficiency, preventing the loss of soil including soil productivity, resilience against flood, drought and storm, averting human-wildlife conflict including through crop and livestock damage, protecting assets of spiritual significance and novel economic opportunities such as from ecotourism. In wider work globally on the conservation or restoration of wetlands and other functional habitats, the language of ecosystem services – the benefits that ecosystems provide to humanity – is valuable for contextualising the importance of protecting or enhancing wetland integrity and functioning in terms that are understood and appreciated by diverse stakeholder groups. These include natural regulation of the extremes of flooding and drought events as well as storm damage, improving the quality of urban environments by providing noise and visual buffering and access to 'green' and 'blue' spaces for residents, providing recreational opportunities and contact with calming, spiritual and educational places, protecting water resources and harvestable wildlife and many more benefits besides.

Shared visions need to be founded on far more robust foundations than theory alone or of altruism, which is at best a volatile commodity and particularly so in depressed markets. They ideally also need to be co-created in terms meaningful to all in society. They need to be multifaceted as well as graphic, expressed in practical and tangible terms relevant and aspirational to the choices that everyone makes, individually and as part of families, communities, companies and other organisations, regions and nations, as all are contributory to societal transition towards sustainability.

Visions can unite us on the journey on which all in society are embarked together. We will all either suffer or benefit from the decisions and actions, or inactions, that we undertake or implicitly sanction. A wise course is to develop and share a graphic and relatable vision of where we want to end up. However simplistically we might subdivide society into apparently discrete sectors, the reality is that we all share a common destiny within which all of these subdivisions play different but ideally complementary roles. All of us, individually and in the enterprises within which we fit, participate in the societal value chains of material use. Linking up is vital, rather than perpetuating the implicit but false assumption that any unit in this chain can solve material use challenges alone. This includes competing businesses that, in reality, depend upon the same enabling environment for success, co-created with other sectors of society such as those in the policy and regulatory environment and wider influences on the market conditions through which materials flow. There is no sector of society that does not have a role in the transition to sustainable material flows and wider aspects of sustainable development. The objective reality is that all societal sub-divisions are part of a potentially symbiotic continuum of endeavour driving society in a sustainable direction.

7.5 LEADERSHIP FOR SYMBIOTIC PROGRESS

There is an anachronistic assumption that leadership comes from 'the top', however defined. Often, it is assumed that governments set direction from the top down, which the rest of society then follows. In practice, this is not, with rare exceptions, the case. At the political level in democratic societies, leadership rotates on election cycles and new incumbents are generally often naive to the complexities of the policy areas they inherit. Outcomes over the longer term may not be primary considerations under electoral cycles within which quick wins are significant spurs for re-election.

Other people may regard knowledge generators and knowledge providers primarily in the academic community as primary influencers of systemic change. Whilst academic learning certainly has crucial roles to play in stimulating societal change through the generation of new knowledge, critical thinking, input to technological advancement and the training of future experts, it is also true that shelves full of learned papers and reports do not of themselves drive change.

Others might regard activism by the NGO community and others in the voluntary sector as a primary driver of change. To a certain extent, this was manifestly true at the outset of the modern 'environmental movement'. Institutions in the voluntary sector and media can be influential in 'speaking truth to power' as well as wider policy prioritisation through, for example, highlighting environmental harm or irresponsibility and social injustices, with some NGOs establishing themselves as incubators of novel solutions often in partnership with other sectors.

It is though the business sector that actually puts physical form on aspirations. It does so by investing in infrastructure and material innovation, be that blindly or ideally directed by a strategic commitment to progress towards sustainability.

In reality, leadership does not come uniformly from any singular sector on this journey. We have already addressed examples of elements of leadership from across all societal sectors. Optimally, this will lead towards a more joined-up approach, albeit these best examples currently represent 'baby steps' towards a more integrated, cross-sectoral approach to sustainability challenges.

Though focused primarily on the conduct of business, the institutional economics model of Williamson[65] addresses interactions with other societal sectors through four levels. Resource allocation including material throughputs and employment outcomes in a business enterprise (Level 2) is influenced both by embeddedness of values in society including amongst consumers and NGOs focusing on societal concerns (Level 1) as well as by corporate governance, for example, through Board and senior staff direction (Level 3) that is in turn influenced by the institution environment (Level 4). This is an economic framing of the fact that the regulatory environment, corporate values and the expectations and 'licence to operate' granted by society impinge on resource use and trading practices. In other words, resource throughput and potential sustainable progress is a product not of individual enterprises alone, but an emergent property of interactions with all sectors of society. Significant social dimensions interwoven within the framework of industrial ecology are also recognised by Graedel and Allenby, with strong parallels with Williamson's model in recognising the interdependent facets of: social equity and environmental justice; consumer behaviour and lifestyles; governance and policy; cultural values and worldviews; and community engagement and participation.[66] All players in society have a role in influencing new thinking and support for innovation of materials, processes and infrastructure, providing the necessary confidence for investment to realise sustainable aspirations.

7.5.1 Leadership from the Business Sector

As tomorrow's operating environment for society and business will inevitably be constrained by sustainability pressures, wise businesses will seek to consider sustainability principles in innovation of material production and use in product design with a view to whole life cycle sustainability. This is where future profit will stem from or, in absentia, where future liabilities may arise.

Much has been written about the breadth of benefits to business of taking a leadership stance in terms of commitment to sustainable development. Advantages stem from supply chain security, including controlling costs and a proactive approach to averting future shortages and liabilities. Then there is the building of trust with investors, customers and regulators as well as enhancing employee loyalty by better connecting with their value systems. A strategic commitment to sustainable development is also likely to pre-empt inevitably tightening regulations as novel issues come under statutory scrutiny. Financial as well as wider benefits also clearly stem from reducing pressures on scarce and contested water, forest, mineral and other resources. And then, of course, we have a deepening global interest in the attainment of carbon neutrality in the face of the climate emergency and arresting degradation of nature in the light of the current biodiversity crisis. Transparent corporate commitment to sustainable goals also engenders confidence in staff, shareholders and other investors as well as partners in other sectors.

There is also a necessary recognition of the 'missing half' of sustainable development – business advantages stemming from better meeting needs as a complement to the necessity to tackle the 'bad stuff' – signalling a change in business paradigms, for which the evolution of more enabling policy and market environments, targeting of NGO pressure and systemically framed knowledge are necessary partners.

Businesses have the choice of investing wisely in material innovations and patterns of use that yield continuing profit through more efficiently and safely meeting human needs in an inevitably changing future or, alternatively, continuing to invent and invest blindly with all the jeopardy that this entails. Whichever track enterprises take, be that by intent or by default, they can be certain that the material means by which societal needs are met into the future cannot be a simple extension of the ways that this has been achieved up until today, however more eco-efficient they may be.

The case for rethinking business and wider resource use models to refocus on the meeting of changing needs on a profitable basis has been thoroughly exercised in previous pages, comprising a more enlightened and robust approach than merely pursuing immediate maximisation of profit. In doing so, we have come across examples of businesses deciding to collaborate for sustainable progress and reaching out to other sectors to achieve a vision.

Leading businesses are now increasingly adopting mission-based approaches with associated tangible targets, linked to improvements and innovations to address societal and environmental challenges. This mission-led approach creates space for innovation, stimulating 'blue skies' thinking about novel approaches required to achieve visionary aspirations rather than simply framed by reacting to current issues of concern and regulatory minima with all the hazards that this entails. A further co-benefit of corporate focus on desirable societal end goals is the motivation of and retention of employees.[67] It also has the potential to stimulate customer loyalty through awareness that corporate goals are infused with a sense of purpose in tangibly addressing broader societal objectives. Market differentiation can in turn lead to greater synergy along value chains with other companies that share, or at least anticipate, customer and/or regulatory pressures to address the same broad alignment with progress with sustainability principles.

Clearly, setting ambitious targets leading stepwise towards often distant goals creates pressure on businesses to deliver. However, all businesses already set targets of one type or another, whether linked to societal goals or of a more inward-looking nature. The way in which targets and supporting metrics are framed is also important, as they need to address progressive steps towards long-term outcomes rather than reflecting short-term thinking or simple avoidance of current threats. Targets must also be realistic about what is attainable in the short and medium terms as businesses must remain profitable not only to survive but to continue to deliver progressive innovation.

A number of practical current examples demonstrate corporate missions to drive progress towards clearly stated sustainability-relevant goals. One is the *Sustainable Living Plan* published by the multinational domestic product company Unilever in 2010 with goals for 2020, setting out a mission and associated targets to drive innovation to tackle social inequality, waste reduction and the climate crisis.[68] Another is Tesla's stated mission to accelerate the world's transition to sustainable energy by building products designed to replace some of the planet's biggest polluters, against which Tesla claimed that its customers avoided releasing over 20 million metric tons of CO_2e into the atmosphere in 2023.[69] The stated mission of UK-based renewable energy company Ecotricity is a society-wide intent to "*...end fossil fuels: Let's replace fossil fuels with green electricity and green gas*".[70] At a broader material sector level, Plastics Europe – a pan-European association of plastics manufacturers with close to 100 members producing over 90% of all polymers across Europe – published *The Plastics Transition* in 2023 recognising the sector's roles in meeting the EU's net-zero and circularity objectives, and setting out a roadmap with milestones to 2030 and associated indicators to monitor progress and transparently report progress made by its members every two years.[71]

A mission-led approach in business also requires ostensibly competing enterprises to work together to influence the operating environment, collaborating to promote and reward sustainable innovation in the pursuit of increasing sustainability. The VinylPlus® voluntary commitment to sustainable development across the European PVC value chain has been an outstanding example of key players – including ostensibly competing additive and resin manufacturers, converters in markets for diverse product groups, recyclers and other players from right along the European vinyl value chain – combining forces and investments to address strategic challenges that can only be met through collaboration. Five founding challenges for the PVC value chain, developed from backcasting using The Natural Step science-based tools published in 2000,[72] still underpin audited targets and drive further innovation and investment towards a shared vision of the sustainable use of PVC.

High recycling rates of paper, glass, gold, copper and other materials have been achieved through collaboration and co-investment along whole value chains, linking with waste handlers and wider societal sectors. Businesses have joined hands and investments with NGOs and other voluntary community groups, and in some cases with regulators in effective supply chains stewardship schemes (FSC, MSC and Rainforest Alliance to name just a few examples expanded in preceding pages) as well as take-back and recycling schemes (previously cited examples include post-consumer cotton, batteries and printer cartridges). Kerbside recycling by municipalities is part of these initiatives contributing to cyclic resource use.

The pursuit of strategic societal sustainability-relevant goals, such as achievement of the SDGs, 'Net Zero' or the circular economy, requires a holistic and innovative approach that genuinely links the energies, creativity and investment of innovators in business with the influences of policymaking and enforcement. Traditional 'bottom-up' regulation and enforcement mechanisms remain essential to set a baseline and to ensure that the laggards and recalcitrant companies do not undermine the bold intent and outcompete those committed to progress through pricing that externalises irresponsible behaviours. However, the 'bottom-up' process alone cannot drive strategic progress, tending instead to drive reactive measures to avoid non-compliance. There is a need for a partnership approach in which regulatory measures support innovative businesses through tracking audited progress with agreed performance improvement targets leading towards longer-term consensual goals, empowering organisations to seek greater security and profitability through innovation and improvement.

7.5.2 Leadership from the Policy and Regulatory Sector

The roots of the policy and regulatory sector were initially reactive, responding to gross environmental, public health and inequitable offences as they rose in popular concern. Leadership in the regulation of chemicals was then framed as setting a common baseline of emissions, permitted substances and environmental quality thresholds. However, the policy and regulatory sector has significant roles to play not merely in establishing a *de minimus* level of performance but in stimulating progress towards sustainability, and this is enshrined not only in multinational agreements such as under the Brundtland Commission but also in domestic legislation in many territories explicitly requiring regulatory action in pursuit of sustainable development.

Rhetorical commitments to sustainable development in government and regulatory circles have though often resolved into little more than eco-efficiency measures that fail to challenge deeply entrenched habits, assumptions and vested interests. This is a facet of regulatory approaches remaining blinkered by historic 'bottom-up' regulatory practices, rather than recognising the qualitative differences required to stimulate progress with sustainable development. Whilst ensuring a baseline level of minimum environmental and social performance remains important for averting poisoning and dispossessing people and harming the ecosystems upon which they depend, and on dealing with resistant organisations, a bottom-up approach alone is not an effective means to drive the scale of paradigmatic change demanded by the challenges now facing society.

There is a clear need to reappraise the purpose of regulation as a driver of progression towards sustainability. A new model of regulation necessarily entails engaging with the 'missing half' of sustainable development: stimulus to accelerate innovation towards safer and more efficient ways of meeting needs now and into an inevitably different future, as agreed consensually at a global scale in the 1980s.

Strategic signals about goals form part of this enabling approach, providing a focal point in the future from which to backcast. But further regulatory evolution is then required to work synergistically with other sectors of society, particularly business, recognising that innovation happens across the whole of society. True partnership between business and regulatory sectors is fundamental, co-creating visionary but scientifically grounded targets that give businesses freedom to innovate and to report on audited targets that build progressively towards agreed goals. Regulatory approaches can then shift emphasis to a symbiotic approach, ideally also in collaboration with knowledge providers as well as with support for a strategic approach from campaigning by the NGO sector, to co-create and agree on sustainability-relevant goals and associated progressive targets. Compliance then shifts from a bottom-up approach alone to one based on the assessment of audited progress reports linked to clear, agreed longer-term aspirations. Regulators can also act in an advisory capacity to help shape, support and potentially co-fund progressive steps towards these consensual goals.

An important part of this is a shift that also improves upon an over-simplistic hazard-based approach that, whilst easy to assess and enforce, can be misleading about actual risks. Missed opportunities resulting from a narrowly hazard-based approach include, for example, overlooking the values provided by presumed 'bad' materials that, in reality, are wholly consumed in production of compounds and products within tightly controlled environments with no consequent environmental and human exposure and hence risk. Risks are also eliminated if substances are immobile within compounds and are then recaptured in technical recycling loops, consistent with commitments to work towards circularity in the economy and in material flows through society. This shift from hazard-based to risk-based assessment within the context of whole life cycles represents an intelligent evolution of regulation. It is an evolution that may most efficiently be achieved by putting the onus on innovating businesses to demonstrate acceptable risk management. Effective models wherein businesses prepare and submit evidence about chemicals for regulatory scrutiny are already established. For example, in Europe, businesses or business consortia are required to develop dossiers of evidence about substances under the REACH process. At the product level, preparation of Environmental Product Declarations (EPDs) by businesses or business consortia also occurs in Europe under the *Construction Products Regulation*[73] mandating the use of EPDs for certain construction products, particularly those with significant environmental impacts with individual European countries also developing their own national programs or regulations promoting or requiring EPDs for various product categories. EPDs are also widely used in Australia as well as in Canada, especially in the building and construction sector and, though not mandated, they are also gaining traction in the US, Japan and China. Were these substance-specific and product-level assessments set within whole-life risk contexts rather than narrowly focused on potential hazard, they could serve the purpose of bridging business and regulation in a more symbiotic way. At the value chain level, VinylPlus brings the whole EU+ PVC chain together around goals relating to industry-wide voluntary commitments, publishing annual reports that document audited progress with associated targets. Annual reports produced by VinylPlus could then constitute the business half of a symbiotic relationship were regulators to develop an appetite for stimulating and supporting innovative progress towards sustainability, breaking away from an anachronistic focus that overlooks positive contributions to meeting needs with appropriate risk assessment.

Grants and regulatory easements also represent potential supportive contributions for sustainable progress from the government and regulatory sector. They could, for example, be applied to promoting take-back and recycling schemes, novel industrial processes, or investment in the generation of renewable energy, charging infrastructure for electric vehicles or recycling facilities. Further economic influence can be brought to bear through the exercise of public procurement, as, for example, the *Recycled First* policy of Sustainability Victoria (addressed previously in this chapter) that favours materials and products with recycled content in infrastructure and other schemes promoting the Australian state of Victoria's circular economy strategy, thereby supporting investment in recycling and innovation of recycled products that can also find new markets outside of state projects.

Across the world today, there are emergent innovative strategies promoted by governments relating, for example, to aspirations towards circular resource use and achievement of 'Net Zero' emissions of climate-active gases. These strategies provide other sectors with a 'direction of travel' and confidence to plan and invest. These visions have power when collectively conceived, gaining traction when backed up by realistic and measurable targets representing milestones towards their achievement. These targets are necessarily stepwise in terms of what is achievable in short- and medium-term timeframes. They also serve to hold policymakers to account. A balance has to be struck, as setting targets that are overly rigid can stifle creativity, dissuade collaboration and achieve only narrow outcomes whilst overlooking wider ramifications. These examples of potentially societally shared visions are currently just sporadic 'green shoots' rather than fully reconstructed regulatory approaches, though they do signal directions in which regulatory thinking, instruments and their implementation need to evolve if they are to work symbiotically to accelerate societal progress towards sustainability.

At a global scale, the 17 United Nations SDGs are as close as the world has ever come to a consensus about desirable end goals for 2030. Missions and associated targets though can and need to be downscaled to address challenges at national or regional scales. As one national-scale example, the new Labour Government in the UK committed to being 'mission-driven' in 2024, focusing its efforts on five key areas: growth, the National Health Service, clean energy, safer streets and opportunity.[74] Downscaling is also important to address and mobilise support and collaboration towards discrete focal challenges at appropriate scales, such as sustainable material use both within and across specific product sectors, food production and distribution, catchment management or other societal challenges.

There is an emerging trend in governments around the world towards adopting a mission-based approach, supported by tangible and often ambitious targets, as a strategic means to tackle complex societal challenges. Establishing a mission makes sense for addressing complex challenges for which no obvious and linear pathways to attainment are evident, but for which it is therefore necessary to galvanise collaboration and collective action, innovation and investment towards the achievement of clear, ideally shared goals. This applies to the challenge of joining up across the frequently limited focus of government departments but, more significantly, also connecting energies and investments across societal sectors. The term 'moon shot' is often used to describe this mission-led approach, so-named after US President

John F. Kennedy's stated goal of landing a man on the moon by the end of the 1960s spurring substantial technological advancements and cooperation across America and with international partners to make this possible. Progress with mission-oriented innovation policies (MOIPs) is currently hesitant. A 2024 report by the Organisation for Economic Co-operation and Development (OECD) found that, although many countries have adopted MOIPs, evidence of impact from the broad and ambitious objectives of 101 'net zero missions' and 17 in-depth case studies remained largely anecdotal.[75] This was largely because strategies outlined in MOIPs "*...often lack clear focus, measurable targets, and systematic monitoring*" with shortfalls in dedicated, multiannual funding and new financial mechanisms to attract private sector investment leading to insufficient engagement with private resources. These findings do not undermine the promise of MOIPs. They do though emphasise the "*...need for a new generation of missions that more closely integrate a broader array of public and private actors and resources to drive transformative change and achieve national net zero targets*". Synergy of aspirations with effective support mechanisms across societal sectors, including incentives and reform of regulatory requirements, is of fundamental importance to provide tangible and beneficial stimulants for material progress towards mission-led policy aspirations.

Governments and regulators also have roles to play in clearing the pathway towards consensual goals through revisiting legacy regulations at all scales from the national to the intergovernmental. Review is necessary to ensure that the laudable purposes of these agreements are respected, but that anachronistic language and framing do not create barriers to subsequently emerging strategic priorities such as promoting circularity, 'Net Zero' and just transition towards efficient and safe use of materials. The annihilation of the EU's 'circular economy' goal by the immediate expectation of achievement of the goal of the partner 'clean chemistry' aspiration has been outlined previously as an example of a conflict that can be resolved through a strategic approach over the longer term (see Section 4.2.6). It is also necessary to challenge and revise the framing or implementation of instruments that unhelpfully classify by-products from industrial processes, or post-consumer or post-industrial materials, as 'wastes' thereby creating legal obstacles or completely preventing beneficial reuse by other industries consistent with the 'industrial ecosystem' model and circular economy aspirations to which nations are ostensibly committed. Problems with the outmoded language of the 1992 Basel Convention, which classifies many substances as 'wastes' or 'hazardous wastes' when they cross borders, have been outlined in Chapter 5. A further consideration for regulators and legislators is to ensure that the ways that anti-competition rules are applied do not prevent potentially competing businesses co-investing and collaborating to accelerate sustainable change.

A pertinent question to ask when reviewing policies, regulations and subsidies across all spheres of societal interest, including those pertaining to the use of materials, is *"What is the destination to which this instrument is aiming?"* Does it have a clear goal, such as achievement of circular resource use, 'Net Zero' climate-active emissions, biodiversity-positive outcomes or distributional equity? Alternatively, does it create an obstacle to what we now recognise as societal priorities? Often, we may find that the instrument under review is aimed in one way or another at being less harmful, or in other words addressing one half of the sustainability challenge, but without the empowering and directional intent of the 'missing half' of stimulating positive progress towards a clear end goal. We have seen many examples throughout these pages of well-intentioned strategies and policies that, without recontextualising within priorities that have emerged subsequently, can become blockers to sustainable progress including: unrealistic immediate expectation of the European 'Clean chemistry' aspiration effectively driving linear resource use and killing off the goal of circular resource use; rigorous implementation of the outmoded language of the 1992 Basel Convention blocking beneficial reuse of industrial by-products and post-consumer materials; and continued focus on potential hazard overlooking actual life cycle risks and mitigation strategies. These gaps represent opportunities to refine the framing or implementation of statutory and strategy mechanisms to better serve progress towards consensually agreed commitments to meet human and environmental needs now and into the future.

7.5.3 Leadership from the Knowledge Generation Sector

In the knowledge generation sector, the *Ending Plastic Waste* mission developed by CSIRO, Australia's national science agency, stands as an exemplar of focus on a consensual vision. CSIRO has created a widely supported vision of ending plastic waste, also graphically illustrating potential contributions from different sectors – industry, research and government – necessary to make concerted progress towards attainment of the uniting vision through reimagining societal use of plastics and transforming plastic waste into an economic commodity amongst a range of other innovations. Means for attainment are not prescriptive but set a shared direction based on consensual cross-sectoral agreements of where we need to end up.

More broadly, a vast body of learned studies help articulate the benefits of sustainable approaches and means for their stepwise attainment. These ultimately need to be integrated into clear visions and multi-sectoral contributions in the manner of the CSIRO vision. To achieve this, there is a need to more deeply embed systems perspectives into research or to invest in outreach and science communication services to contextualise still-important disciplinary-bound research in terms of systemic and societally relatable implications along with associated opportunities and risks.

The research and knowledge-providing sector can also support government, business and the voluntary sector in the development of robustly science-based visions and relevant targets, as well as metrics to measure and audit progress towards them. It can also recommend or develop social processes to support synergy and co-creation of goals across societal sectors, as well as tools to resolve conflicts between them. It also clearly has a role in training future professionals in systemic thinking, whatever the discipline in which they are trained, such that they are better equipped to contribute to sustainable progress.

7.5.4 Leadership from the Voluntary Sector

For decades, non-governmental organisations within the voluntary sector have been highly effective in aggregating wider and emergent societal concerns and campaigning to raise awareness about issues that may formerly have been overlooked by public, business and government. Awareness-raising about neglected or ignored aspects of sustainability is still necessary, including critique of broken promises and flaccid commitments. However, many NGOs have moved into the solutions space, working collaboratively with business and government sectors to establish visions and stepwise progress towards them. The NGO *Forum for the Future*[76] has been particularly articulate in working with major businesses, governments and civil society at all scales to address and inform progress towards many sustainability challenges. This includes the role that *Forum* formerly played in hosting the UK office of The Natural Step, which was the innovator of the 'Five TNS Sustainability Challenges for PVC' in 2000 that have underpinned a great deal of collaborative progress along the European PVC value chain for the past quarter-century as well as exerting wider global influence.

As we have seen, NGOs and other voluntary groups have also collaborated with multi-partner initiatives. One example is the collaborative role played by the NGO Greenpeace in the development and establishment of the FSC. The WWF has joined the RTRS to exert influence and is also a partner in the development of the ASC, and in co-convening the *Business Coalition for a Global Plastics Treaty*. I must put my hand up here too as a 'one-man NGO' beyond my work on the vision of sustainable PVC when with The Natural Step, as I have also contributed to positions and tools for the Ramsar Convention, and subsequently developed a practical vision in the material use sector relating to the 'Level playing field' concept[77] that is beginning to shape how people in the industry think about optimal material selection and use. This new book also creates a vision and pathway towards the wider societal dynamics and symbiosis required to bring that 'level playing field' to life.

Just as the regulatory sector has a role in revisiting former policies and protocols that may now constitute obstacles towards subsequently emerging priorities, so too the NGO sector is advised to revisit former hard-line positions. These may include, for example, former branding of 'bad' or 'good' materials in the absence of life cycle risk assessment that may now create obstacles to resource recovery and recycling, or optimally efficient and safe meeting of needs. Under a symbiotic approach united around future desirable destinations, a more supportive attitude to progressive and currently attainable steps would be welcome: we have seen in the example of unrealistic expectations of immediate attainment of the strategic goals of the EU 'Clean chemistry' and 'Circular economy' strategies how ignoring the wisdom of Voltaire – "*The best is the enemy of the good*" – can disincentivise or wholly derail sustainable progress.

A mission-led approach focused on desirable and consensual societal end goals, encouraging and recognising progressive steps towards them, would better empower the voluntary sector as a more effective partner for the promotion of sustainable change.

7.5.5 Leadership from Marketing

Whilst marketing companies are part of the business sector, and direct business communications also form part of this picture, marketing activities are drawn out here as they have significant roles to play in influencing consumption habits. Whilst it has a long track record of stimulating over-consumption, particularly of unnecessary 'false satisfiers' (as defined by Manfred Max-Neef), marketing has the power to modify its messaging to the wider public about the benefits that can result from more sustainable habits and products. Companies commissioning advertisements share this responsibility.

If 'green consumerism' is to thrive and change market norms, marketing messages to general consumers matters. So too does promotion of products to bulk trade markets. A shift in emphasis is required to 'sell' messages about the virtues and self-benefits of long-term durability, low maintenance requirements, reduction in associated costs and reduced potential liabilities stemming from responsible sourcing and high potential for recovery and recycling. This is both an opportunity and, if commitments to a sustainable pathway of development are to be believed, an obligation.

There is a handshake between marketing and public policy around green claims and 'greenwashing', backing up a call by António Guterres, UN Secretary-General, that "*We must have zero tolerance for net-zero greenwashing*". In the European Union, the 2024 Empowering Consumers Directive (Directive 2024/825)[78] aims to better protect consumers against unfair practices, and a proposed Green Claims Directive[79] passed its first reading in March 2024 to improve the legal certainty with respect to environmental claims. Elsewhere, the 2010 *Australian Consumer Law*[80] prohibits making false or misleading representations about goods or services and, in Singapore, the *Consumer Protection (Fair Trading) Act 2003*[81] applies to green claims within its wider aim of protecting consumers against unfair practices.

The bulk of contemporary marketing by specialists and companies is poorly oriented towards, or indeed entirely inconsistent with, stated commitments to sustainable development. A cross-societal mission-led approach requires more responsible messages promoted in marketing materials, grounded in robust evidence and principles relevant to sustainable development and expressed in terms that articulate benefits to consumers.

7.5.6 Assistive Digital Technologies

Section 6.4 outlined aspects of the rapid acceleration and permeation of the digital world into all facets of human activity, not least including into value chains of which chemical usage constitutes a significant part. Digital tools to facilitate collaboration across societal sectors, including value chain auditing and platforms for consensus-building as well as the sharing of knowledge, can play key roles in linking partners across society to accelerate progress towards the sustainable use of materials. They can also transparently reveal broader implications arising from decisions relating to the innovation, use and fate of materials, enabling greater collaboration between partners in value chains.

The potential for the use of current and still-emerging digital assistive technology is a broad topic. However, clearly, enhanced modelling, visualisation and the potential use of blockchain and artificial intelligence all offer substantial potential for enhancing transparency and for connecting societal sectors to make progress towards the beneficial and sustainable use of substances, contributing to wider societal goals such as circular economy, reduced climate-active emissions and greater accountability for environmental and ethical footprints.

NOTES

1 Haberl, H., Fischer-Kowalski, M., Krausmann, F., Martinez-Alier, J. and Winiwarter, V. (2011). A socio-metabolic transition towards sustainability? Challenges for another great transformation. *Sustainable Development*, 19(1), pp. 1–14. https://doi.org/10.1002/sd.410.

2 Everard, M. (2000). Aquatic ecology, economy and society: the place of aquatic ecology in the sustainability agenda. *Freshwater Forum*, 13, pp. 31–46.

3 Robèrt, K.-H. (2022). *The Natural Step Story: Seeding a Quiet Revolution*. New Society Publishers, Gabriola Island.

4   Packard, J. (2019). *Monterey Bay Aquarium: The First 35 Years.* Monterey Bay Aquarium. [Online.] https://www.montereybayaquarium.org/globalassets/mba/pdf/join-and-give/aquarium-the-first-35-years.pdf, accessed 13 October 2024.

5   Packard Foundation. (2024). *How the Monterey Bay Aquarium is Shaping the Future of Ocean Conservation.* Packard Foundation. [Online.] https://www.packard.org/insights/grantee-story/how-the-monterey-bay-aquarium-is-shaping-the-future-of-ocean-conservation/, accessed 15 November 2024.

6   Gates Foundation. (2025). *Partners of Human Potential.* Gates Foundation. [Online.] https://www.gatesfoundation.org/, accessed 17 October 2024.

7   FSC United Kingdom. (2023). *What is FSC?* Forest Stewardship Council UK. [Online.] https://uk.fsc.org/what-is-fsc, accessed 20 April 2024.

8   MSC. (2024). *Our Strategy.* [Online.] https://www.msc.org/about-the-msc/our-strategy, accessed 20 April 2024.

9   ASC. (204). *Setting the Standard for Seafood.* Aquaculture Stewardship Council (ASC). [Online.] https://asc-aqua.org/, accessed 18 October 2024.

10 Rainforest Alliance. (2024). *What does "Rainforest Alliance Certified" Mean?* Rainforest Alliance. [Online.] https://www.rainforest-alliance.org/, accessed 18 October 2024.

11 Red Tractor. (2024). *Welcome.* Red Tractor. [Online.] https://redtractor.org.uk/, accessed 18 October 2024.

12 RSPO. (204). *A Global Partnership to Make Palm Oil Sustainable.* Roundtable on Sustainable Palm Oil (RSPO). [Online.] https://rspo.org/, accessed 18 October 2024.

13 RTRS. (2024). *What are the Benefits of RTRS Certification?* Round Table on Responsible Soy (RTRS). [Online.] https://responsiblesoy.org/certificacion?lang=en, accessed 23 September 2024.

14 Responsible Minerals Initiative. (n.d.) *Conflict Minerals Reporting Template.* Responsible Minerals Initiative. [Online.] https://www.responsiblemineralsinitiative.org/reporting-templates/cmrt/, accessed 12 October 2024.

15 Parra-Paitan, C., zu Ermgassen, E.K.H.J., Meyfroidt, P. and Verburg, P.H. (2023). Large gaps in voluntary sustainability commitments covering the global cocoa trade. *Global Environmental Change*, 81, p. 102696. https://doi.org/10.1016/j.gloenvcha.2023.102696.

16 Unilever. (2021). *Unilever Commits to Help Build a More Inclusive Society.* Unilever, 21 January 20231. [Online.] https://www.unilever.com/news/press-and-media/press-releases/2021/unilever-commits-to-help-build-a-more-inclusive-society/, accessed 21 October 2024.

17 Better Cotton. (2024). *What is Better Cotton?* Better Cotton. [Online.] https://bettercotton.org/, accessed 12 October 2024.

18 ICCA. (2023). *Responsible Care®.* International Council of Chemical Associations (ICCA). [Online.] https://icca-chem.org/focus/responsible-care/, accessed 14 May 2023.

19 King, A.A. and Lenox, M.J. (2000). Industry self-regulation without sanctions: the chemical industry's responsible care program. *The Academy of Management Journal*, 43(4), pp. 698–716. https://www.researchgate.net/publication/324996428_Industry_Self-Regulation_Without_Sanctions_The_Chemical_Industry's_Responsible_Care_Program.

20 VinylPlus. (2020). *VinylPlus®: Steering the PVC Industry towards the Circular Economy.* VinylPlus, Brussels. [Online.] https://www.vinylplus.eu/news/vinylplus-steering-the-pvc-industry-towards-the-circular-economy/, accessed 18 October 2024.

21 Vinyl Council Australia. (2024). *PVC Stewardship*. Vinyl Council Australia. [Online.] https://www.vinyl.org.au/sustainability/stewardship, accessed 15 November 2024.

22 Office for Product Safety and Standards. (2024). *Consumer Products: Recycling Batteries and Electrical Waste*. Office for Product Safety and Standards, 05 March 2024. https://www.gov.uk/guidance/consumer-products-recycling-batteries-and-electrical-waste, accessed 18 October 2024.

23 Kuyichi. (2024). *Post-Consumer Recycled Cotton; Turn Your Throw-Away into Something New*. Kuyichci. [Online.] https://kuyichi.com/blog/post-consumer-recycled-cotton-turn-your-throw-away-into-something-new/, accessed 21 October 2024.

24 Johnson, Z., Mao, H., Lefebvre, S. and Ganesh, J. (2019). Good guys can finish first: how brand reputation affects extension evaluations. *Journal of Consumer Psychology*, 29(4), pp. 565–583. https://doi.org/10.1002/jcpy.1109.

25 Menghwar, P.S. and Daood, A. (2021). Creating shared value: a systematic review, synthesis and integrative perspective. *International Journal of Management Review*, 23(4), pp. 466–485. https://doi.org/10.1111/ijmr.12252.

26 Honglei Mu, H. and Lee, Y. (2023). Greenwashing in corporate social responsibility: a dual-faceted analysis of its impact on employee trust and identification. *Sustainability*, 15(22), p. 15693. https://doi.org/10.3390/su152215693.

27 Paumgarten, N. (2016). Patagonia's Philosopher-King. *The New Yorker*, 12 September 2016.

28 Economics Observatory. (2024). *Mandatory corporate reporting on sustainability: what is the likely impact?* Economics Observatory. [Online.] https://www.economicsobservatory.com/mandatory-corporate-reporting-on-sustainability-what-is-the-likely-impact, accessed 29 October 2024.

29 KPMG. (2024) *Our Insights*. KPMG. [Online.] https://kpmg.com/xx/en/our-insights.html, accessed 29 October 2024.

30 Christensen, H.B., Hail, L. and Leuz, C. (2021). Mandatory CSR and sustainability reporting: economic analysis and literature review. *Review of Accounting Studies*, 26, pp. 1176–1248. https://doi.org/10.1007/s11142-021-09609-5.

31 GRI. (2024). *The Global Leader for Impact Reporting* GRI. [Online.] https://www.globalreporting.org/, accessed 29 October 2024.

32 IFRS. (2024). *IASB Launches Review of the Statement of Cash Flows*. International Financial Reporting Standards Foundation (IFRS). [Online.] https://www.ifrs.org/, accessed 29 October 2024.

33 Gelles, D. (2023). How Environmentally Conscious Investing Became a Target of Conservatives. *The New York Times*, 28 February 2023.

34 United Nations Global Compact. (2013). *Growing into Your Sustainability Commitments: A Roadmap for Impact and Value Creation*. United Nations. [Online.] https://unglobalcompact.org/library/551, accessed 21 October 2024.

35 European Commission. (2022). *Ecodesign for Sustainable Products*. European Commission, Brussels. [Online.] https://commission.europa.eu/energy-climate-change-environment/standards-tools-and-labels/products-labelling-rules-and-requirements/sustainable-products/ecodesign-sustainable-products_en, accessed 14 May 2023.

36 CMS. (2023). *Plastics and Packaging Laws in France*. CMS Law. [Online.] https://cms.law/en/int/expert-guides/plastics-and-packaging-laws/france, accessed 13 May 2023.

37 Parliament of the Commonwealth of Australia. (2024). *Treasury Laws Amendment (Financial Market Infrastructure and Other Measures) Bill 2024.* Parliament of the Commonwealth of Australia. [Online.] https://parlinfo.aph. gov.au/parlInfo/download/legislation/bills/r7176_aspassed/toc_pdf/24042b01. pdf;fileType=application%2Fpdf, accessed 08 November 2024.
38 EU. (2024). *Regulation on Deforestation-Free Products.* European Union (EU). [Online.] https://environment.ec.europa.eu/topics/forests/deforestation/regulation-deforestation-free-products_en, accessed 18 November 224.
39 DOI. (2024). *Global China 2049 Initiative.* ODI. [Online.] https://odi.org/en/about/our-work/global-china-2049-initiative/, accessed 18 October 2024.
40 Pearson, J., Naselaris, T., Holmes, E.A. and Kosslyn, S.M. (2015). Mental imagery: functional mechanisms and clinical applications. *Trends in Cognitive Science*, 19(10), pp. 590–602. https://doi.org/10.1016/j.tics.2015.08.003.
41 Merlo, F. (2021). Vision 2050 Products & Materials Pathway: We Can Make Things, Smarter. *WBCSD Insights*, 16 September 2021. [Online.] https://www.wbcsd.org/news/vision-2050-products-materials-pathway/, accessed 18 October 2024.
42 WBCSD. (2024). *Part Three: Time for a Mindset Shift.* World Business Council for Sustainable Development (WBCSD). [Online.] https://timetotransform.biz/wp-content/uploads/2021/03/WBSCD_Vision2050_Time-for-a-mindset-shift. pdf, accessed 19 October 2024.
43 Everard, M. (2020). *Rebuilding the Earth: Regenerating Our Planet's Life Support Systems for a Sustainable Future.* Palgrave Macmillan, Cham.
44 Costanza, R. (2023). *Addicted to Growth: Societal Therapy for a Sustainable Wellbeing Future.* Earthscan by Routledge, Abingdon. 136pp.
45 German Federal Foreign Office. (n.d.) Energy in Transition - Powering Tomorrow. [Online.] https://energiewende-global.com/en/, accessed 03 May 2025.
46 Friedman, T.L. (2015). Germany, the Green Superpower. *International New York Times*, Thursday 7th May 2015.
47 Amelang, S. (2024). Renewables Cover 56% of Germany's Power Consumption in First Quarter of the Year, Utilities Say. *Clean Energy Wire*, 26 Apr 2024. [Online.] https://www.cleanenergywire.org/news/renewables-cover-56-germanys-power-consumption-first-quarter-year-utilities-say, accessed 18 October 2024.
48 Maliszewska-Nienartowicz, J. (2024). Impact of Russia's invasion of Ukraine on renewable energy development in Germany and Italy. *Utilities Policy*, 87, p. 101731. https://doi.org/10.1016/j.jup.2024.101731.
49 CSIRO. (2024). *Ending Plastic Waste.* CSIRO. [Online.] https://www.csiro.au/en/about/challenges-missions/ending-plastic-waste, accessed 31 October 2024.
50 VinylPlus. (2024). *VinylPlus, Committed to Sustainable Development.* VinylPlus. [Online.] https://www.vinylplus.eu/, accessed 18 October 2024.
51 Everard, M., Monaghan, M. and Ray, D. (2000). *PVC: An Evaluation Using the Natural Step Framework.* The Natural Step, Cheltenham.
52 UNEP. (2024). *Sustainable Consumption and Production Policies.* United Nations Environment Programme (UNEP). [Online.] https://www.unep. org/explore-topics/resource-efficiency/what-we-do/sustainable-consumption-and-production-policies, accessed 18 October 2024.
53 EPA. (2024). *Sustainable Materials Management.* US Environmental Protection Agency (EPA). [Online.] https://www.epa.gov/smm, accessed 19 October 2024.

54 Polystar. (204). *What is Polyethylene Used For? – Polyethylene Plastic Cycle in the Circular Economy*. Polystar. [Online.] https://www.polystarco.com/blog-detail/polyethylene-used-for-polyethylene-plastic-cycle-in-the-circular-economy/, accessed 19 October 2024.

55 Bell, J. (2024). *7 Zero Waste Clothing Brands Moving Circular Fashion Forward*. sustainablejungle.com, 9 September 2024. [Online.] https://www.sustainablejungle.com/zero-waste-clothing-brands/, accessed 21 October 2024.

56 sustainablejungle.com. (2024). *Our Ratings & Reviews Explained*. [Online.] https://www.sustainablejungle.com/our-ratings-explained/, accessed 21 October 2024.

57 CPA. (2024). *Design for Recycling Work Plan: Final Draft – Version 04 March 2020*. Circular Plastics Alliance (CPA). [Online.] https://plasticseurope.org/roadmap-cases/circular-plastics-alliance-cpa/, accessed 21 October 2024.

58 Ellen Macarthur Foundation. (2024). *The Global Commitment: A Common Vision for a Circular Economy for Plastics*. Ellen Macarthur Foundation. [Online.] https://www.ellenmacarthurfoundation.org/global-commitment/overview, accessed 21 October 2024.

59 Business Coalition for a Global Plastics Treaty. (2024). *Business Coalition for a Global Plastics Treaty*. [Online.] https://www.businessforplasticstreaty.org/, accessed 21 October 2024.

60 Interface. (2024). *Evergreen Lease: Flooring Leasing Solution*. Interface. [Online.] https://www.interface.com/AU/en-AU/sustainability/evergreen-lease.html, accessed 21 October 2024.

61 SV. (2020). *Recycled First Policy, March 2020*. Sustainability Victoria (SV). [Online.] https://bigbuild.vic.gov.au/__data/assets/pdf_file/0008/702863/Recycled-First-Policy.pdf, accessed 15 November 2024.

62 European Parliament. (2024). *'Green Claims' Directive: Protecting Consumers from Greenwashing*. European Parliament Think Tank Briefing 11-10-2024. [Online.] https://www.europarl.europa.eu/news/en/agenda/briefing/2024-03-11/8/green-claims-protecting-consumers-from-being-misled, accessed 21 November 2024.

63 Diamond, J. (2011). *Collapse: How Societies Choose to Fail or Survive*. Penguin, London.

64 ONS. (2024). *The National Statistics Socio-Economic Classification (NS-SEC)*. Office for National Statistics (ONS). [Online.] https://www.ons.gov.uk/methodology/classificationsandstandards/otherclassifications/thenationalstatisticssocioeconomicclassificationnssecrebasedonsoc2010, accessed 07 January 2025.

65 Williamson, O.E. (2000) The new institutional economics: taking stock, looking ahead. *Journal of Economic Literature*, 38(3), pp. 595–613. https://doi.org/10.1177/1086026612475068.

66 Graedel, T.E. and Allenby, B. (2010). *Industrial Ecology and Sustainable Engineering*. Pearson, London. 403pp.

67 McKinsey. (2021). *Help Your Employees Find Purpose—or Watch Them Leave*. McKinsey. [Online.] https://www.mckinsey.com/capabilities/people-and-organizational-performance/our-insights/help-your-employees-find-purpose-or-watch-them-leave, accessed 06 December 2024.

68 Unilever. (2020). *Unilever Celebrates 10 Years of the Sustainable Living Plan*. Unilever, 14 May 2020. [Online.] https://www.unilever.co.uk/news/press-releases/2020/unilever-celebrates-10-years-of-the-sustainable-living-plan/, accessed 06 December 2024.

69   Tesla. (2023). *Impact Report 2023: A Sustainable Future is Within Reach.* Tesla. [Online.] https://www.tesla.com/en_gb/impact, accessed 06 December 2024.

70   Ecotricity. (2024). *Our Mission: End Fossil Fuels – Let's Replace Fossil Fuels with Green Electricity and Green Gas.* Ecotricity. [Online.] https://www.ecotricity.co.uk/our-story/our-mission, accesses 08 December 2024.

71   Plastics Europe. (2023). *The Plastics Transition: Our Industry's Roadmap for Plastics in Europe to be Circular and Have Net-Zero Emissions by 2050.* Plastics Europe. [Online.] https://plasticseurope.org/wp-content/uploads/2023/10/PlasticsEurope_Summary_24.10.pdf, accessed 06 December 2024.

72   Everard, M., Monaghan, M. and Ray, D. (2000). *PVC: An Evaluation Using the Natural Step Framework.* The Natural Step, Cheltenham.

73   European Union. (2011). *Regulation (EU) No 305/2011 of the European Parliament and of the Council of 9 March 2011 Laying Down Harmonised Conditions for the Marketing Of Construction Products and Repealing Council Directive 89/106/EEC Text with EEA Relevance.* European Union. [Online.] https://eur-lex.europa.eu/eli/reg/2011/305/oj, accessed 23 January 2025.

74   Institute for Government. (2024). *What does a 'Mission-Driven' Approach to Government Mean and How Can It Be Delivered? How Government could Effectively Organise Itself to Deliver Missions.* Institute for Government, 15 July 2024. [Online.] https://www.instituteforgovernment.org.uk/publication/mission-driven-approach-government#:~:text=The%20new%20government%20has%20committed,energy%2C%20safer%20streets%20and%20opportunity, accessed 06 December 2024.

75   OECD. (2024). *Mission-Oriented Innovation Policies for Net Zero: How Can Countries Implement Missions to Achieve Climate Targets?* OECD Publishing, Paris. https://doi.org/10.1787/5efdbc5c-en.

76   Forum for the Future. (n.d.). *Forum for the Future.* Forum for the Future. [Online.] https://www.forumforthefuture.org/, accessed 01 January 2025.

77   Everard, M. (2024). *Seeking Sustainable Development on a Level Playing Field: A PVC Case Study.* CRC Press, Boca Raton, FL.

78   EU. (2024). *Directive (EU) 2024/825 of the European Parliament and of the Council of 28 February 2024 Amending Directives 2005/29/EC and 2011/83/EU as Regards Empowering Consumers for the Green Transition through Better Protection against Unfair Practices and through Better Information (PE/64/2023/REV/1).* European Union (EU). [Online.] https://eur-lex.europa.eu/legal-content/EN/TXT/?uri=OJ:L_202400825, accessed 01 November 2024.

79   EU. (2024). *Proposal for a DIRECTIVE OF THE EUROPEAN PARLIAMENT AND OF THE COUNCIL on Substantiation and Communication of Explicit Environmental Claims (Green Claims Directive) – COM/2023/166 final).* European Union (EU). [Online.] https://eur-lex.europa.eu/legal-content/EN/TXT/?uri=COM%3A2023%3A0166%3AFIN, accessed 01 November 2024.

80   Australian Government. (2024). *Competition and Consumer Act 2010.* Australian Government. [Online.] https://www.legislation.gov.au/C2004A00109/latest/text, accessed 01 November 2024.

81   Singapore Statutes Online. (2024). *Consumer Protection (Fair Trading) Act 2003.* Singapore Statutes Online. [Online.] https://sso.agc.gov.sg/act/cpfta2003, accessed 01 November 2024.

Realising Symbiotic Value Chains

8

No organism, person, business, corporate sector, life cycle stage, regulator, academic venture or NGO is an island. All are facets of the amorphous yet interconnected whole that we refer to as 'society'. And society is, in turn, fully dependent upon and interdependent with the biosphere within which humanity co-evolved. Our trajectory of development has been supported by, but largely undertaken without regard for, this indivisible synergy with planetary ecosystems, upon which our social metabolism has a profound influence. The legacy of this fragmented trajectory now therefore presents us with major challenges and a pressing need to reorient ourselves collectively onto a pathway of sustainable development.

8.1 A PYRAMIDAL APPROACH TOWARDS SOCIETAL SYMBIOSIS

It is hardly revelatory, not least this far into this book, to recognise that we start from unsustainable norms. In this baseline state, notwithstanding the green shoots we have visited, there is evident friction between and within different principal sectors of society. Business has leaders and laggards with respect to sustainable innovation or resistance to change, with antagonism from both the pioneers and the unwilling about regulatory and non-governmental organisation (NGO) pressures that may inhibit rather than stimulate innovation and confident investment. Regulators would like to push harder though are often still limited in vision to bottom-up approaches, and NGOs as well as other voluntary groups may feel they are agitating in society's best interests but they feel no one is listening. Knowledge-providers exist on a spectrum from a narrow disciplinary focus through to those making systemic connections, including in the

DOI: 10.1201/9781003637875-8

advice they have to give and the students they train, many of these institutions and concerned individuals also frustrated that society at large is ignoring their clear messages about threats and potential solutions. All participants in this game may feel that they are on the side of the angels, but also that their energies are fragmented and not understood or appreciated by other sectors.

This may be an unduly pessimistic summing up, but purposeful coherence across all sectors of society is often hard to find. Problems arise when we are blinkered by today's conflicts and divisions, rather than taking a long-sighted perspective and seeking the outcomes upon which we can all agree regardless of whether they are immediately attainable. But it is that longer view, as a basis for backcasting from consensual visions, that can unite us and remind us that we are all in this together.

Keeping our eyes on agreed destinations on or beyond the horizon better places us to harness and harmonise the agency and energy of all sectors, including their constituent enterprises and individuals, to accelerate progress towards beneficial sustainable goals. Collaborative development of shared visions is important to focus all sectors of society, or at least the far-sighted members of each sector, on the desirability, security and wider benefits of building a sustainable future, also recognising that this is a long-term venture for which stepwise progress is necessary.

Co-created visions, resolving disagreements by looking beyond immediate differences to distant commonly understood and shared 'pole stars' by which to navigate, provide a basis for mutual backcasting from consensual goals towards which all can combine intentions, energies, creativity and investment. It is the intent to work together towards attaining these distant goals that can and needs to bind us and that can help us recognise what sustainable innovation, enabling legislation and subsidies, strategic rather than fragmented campaigning, and the development of systemic knowledge, solutions and skills actually look like.

This can be represented as a pyramid. At its base, the four societal sectors are disconnected and in potential antagonism, though there is the potential for them to converge at the apex under common, co-created visions towards which energies can be synchronised (Figure 8.1). Development of compelling, shared visions cannot be a mere one-off exercise, but rather one that requires regular review and refinement as we learn more together, recognise naiveties in yesterday's framings, and backcast anew to maintain our focus on the attainment of tomorrow's uniting sustainability goals. United, we are strong, progressive and can more rapidly realise a desirable future. Shared 'pole star' visions direct our eyes to the horizon rather than the day-to-day frictions that can divide us, providing goals or strategies to identify optimal resolutions and integrate the contributions of each sector and institution to make these visions real and material.

The apex of the pyramid is the consensual goal of the 'Brundtland definition' of meeting needs now and tomorrow. It will take the form of a range of longer-term, application-specific visions shared by different sectoral interests.

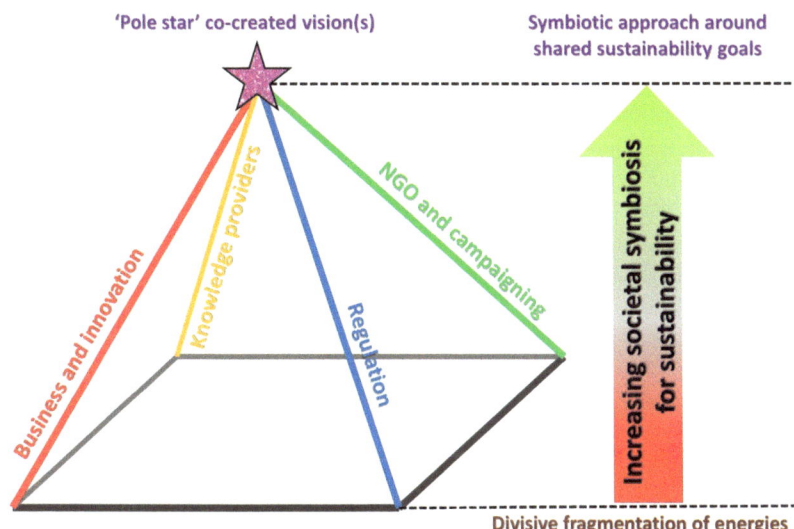

FIGURE 8.1 Representation of progress towards symbiosis across societal sectors, from a divided baseline to coherence around common, shared visions.

A closing consideration about this 'pyramid' model is that some may object to the omission of a spiritual fifth dimension. Representing different belief systems is complex, with evident divisions between some religions, traditions, taboos and other tenets. However, spirituality is far from absent from this model. Rather, it is the 'ghost in the machine' as the moral guidance of belief systems, though in many cases divisive in the world today, can be potent motivators for collaboration in the co-creation and pursuit of desirable goals leading to a better future for all. And, of course, however we go about achieving these goals, it is the use of materials that gives these moral compasses tangible form.

8.2 BACKCASTING FROM A SHARED VISION FOR SUSTAINABILITY

Chapter 7 recognised that, for society to seriously and strategically engage with sustainable development, it is necessary to integrate a suite of principles: (1) a systemic approach to sustainability principles including chemical, physical and wider socio-economic ramifications; (2) considering material use within the whole societal life cycles of products; (3) a precautionary approach founded on strategic goals rather than being derailed by immediate potential hazard; (4) acknowledging that sustainable development is a journey requiring symbiotic collaboration between all societal sectors; and (5) recognising the importance of making and rewarding stepwise progress in the right strategic direction rather than expecting immediate perfect fulfilment of sustainability goals. For this, we need to develop approaches presenting clear, robustly science-based visions for the development of both cross-societal consensus and a basis for backcasting.

The underpinning principles of a robustly science-based framework have been articulated in Chapter 6 when discussing 'Tools for the job'. Chapter 7 summarises a range of initiatives that seek to harness cross-sectoral engagement around clearly articulated, aspirational targets. Figure 8.2 orients these various initiatives on the two axes of 'Basis for backcasting' and 'Cross-sectoral consensus', noting that increasing symbiosis across societal sectors can occur where these two conditions are met (of course, ensuring that they are based on science-based frameworks, which would be a third dimension of the graph).

Traditional bottom-up regulatory approaches driven by government and acting on the minimum performance of business are placed at the bottom-left corner of the graph, as they represent a low basis for backcasting also with low cross-societal consensus (though nonetheless sufficient for acceptance into statute). The UN SDGs are illustratively placed at the other extreme (a high basis both for backcasting and cross-societal consensus), not because they are perfect nor universally understood but because they are as close as the world has yet come to an agreed, cross-sectoral set of desirable goals.

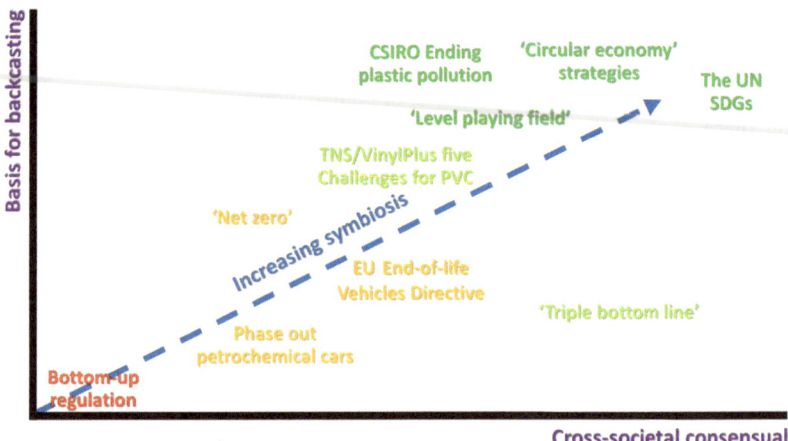

FIGURE 8.2   Illustrative orientation of outcome-based initiatives against the axes of 'Basis for backcasting' and 'Cross-sectoral consensus'.

A range of the initiatives discussed in this book are oriented between these extremes. For example, the 'triple bottom line' articulation of sustainable development might be widely known and consensual (to the right of the X-axis) but implementing it as a basis for backcasting is more elusive (hence low on the Y-axis). Phasing out of petrochemical cars has a degree of societal support but has more of an 'avoid today's problem' framing that may inadvertently drive investment in tomorrow's problems (such as lithium limitation or co-pollutants in electric cars). By contrast, the 'net zero' target sets a broader aspiration that may be less well understood though is open-ended rather than prescriptive about the means for its achievement. The EU's *End-of-Life Vehicles Directive* also sets timelines for percentages of recovery and recycling, providing business with confidence to invest in the necessary infrastructure and subsequent markets for recyclate.

Particularly germane to material use, the aspirational and multi-sectoral goal of CSIRO's *Ending Plastic Waste* mission has already been discussed in terms of its consensual framing around a desirable outcome and its intent to invite collaboration and co-delivery from multiple societal sectors, hence its positioning towards the top right of the illustrative graph. Likewise, circular economy strategies, evident in Europe, Australia, the US and many other parts of the world, set related aspirations that are widely agreed as constituting the basis for backcasting and requiring collaborative innovation across societal sectors. The TNS/VinylPlus approach, substantially structured around five sustainability challenges informed by backcasting from science-based sustainability principles, has substantial take-up across the polyvinyl chloride (PVC) value chain across Europe as well as influencing thinking more widely across the world, albeit that engagement of the regulatory and some parts of the voluntary sector has been less than wholehearted despite significant, audited prog ress towards voluntary targets by multiple businesses. The 'material-blind' *Level Playing Field* approach[1] has also been influential at least in some business sectors in Europe and beyond in terms of the application of common sustainability principles to material choice and patterns of use to optimally service needs in the safest and most efficient way in different applications.

Whilst reviewing the orientation of a selection of today's initiatives with respect to a symbiotic, cross-sectoral approach to the sustainable use of materials, two factors are clearly evident. Firstly, none is perfect, though all nonetheless represent progress towards the kind of novel symbiotic approach that we will have to co-create. Secondly, there remain yawning gaps in terms of integrating the interests of societal sectors to accelerate progress towards clearly articulated sustainability goals. Where we are today is very much a 'work in progress', and one that needs to be accelerated massively if we are to grasp the opportunities of a qualitatively different future whilst avoiding the worst excesses of harm should we fail to change our fragmented ways.

8.3 BRINGING THE PYRAMID TO LIFE

In addition to being a useful conceptual framing, the pyramid model is also a practical tool for understanding where one is in terms of 'pole star' challenges and how one needs to progress together with partners across society to make tangible headway towards consensual and aspirational goals.

8.3.1 The 'Boxing Ring'

The base square of the pyramid – where we largely start today – relates to the things that divide us, with players in these societal sectors perceived or positioned as opposing forces at the corners of what can feel like the square base of a boxing ring. With exceptions, the predominant perception of many institutions within societal sectors, of themselves and of those in other sectors, is one of antagonism.

Regulators often mistrust industry with a focus on minimal performance standards. Regulatory focus consequently acts by majority in the 'boxing ring', driven by a faith that incremental progress from the bottom-up enforcement can move society in the direction of sustainability. As we have observed through many examples of regrettable substitution, as well as energy efficiency investments that serve only to lock in processes and products that are themselves priorities for change, driving progress from the bottom upwards is not a strategic approach. Sole or overwhelming reliance on anachronistic bottom-up regulatory approaches may be important for ensuring minimum levels of acceptable performance by all players within business sectors, but it does not essentially challenge paradigms of resource consumption and use that underpin today's major sustainability concerns.

Many NGOs today start with an assumption of corporate irresponsibility, defining themselves by an oppositional stance and often failing to recognise and encourage incrementally progressive steps on the basis that they fall short of the generally immediately unattainable goal of full sustainability. The same attitude is commonly expressed by NGOs towards the adequacy of regulation and enforcement.

In light of these conflicts, it is hardly surprising that many businesses, reeling under punches within the four-sided 'boxing ring', revert to defensive behaviours or simply seek to be 'less bad' rather than feeling empowered to innovate and invest for greater sustainability.

Divisions in wider societal perceptions, feeding through to funding, can also limit knowledge-providers and educational focus to issues of detail rather than a systemic overview, thereby often failing to address what is required to best inform progress towards longer-term goals.

8.3.2 Guidance from 'Pole Star' Visions

The apex of the pyramid represents the things that unite us in terms of meeting human needs in the safest and most efficient way, acknowledging that this is a longer-term destination attainable only through combined efforts around common challenges. The apex is the place from which to backcast, working out how to combine energies and ingenuity to innovate together for practical stepwise progression towards a better future for all.

For each sector, there is a polarity from what is needed in the short term to what is required to attain a different longer-term future. Commitments to end goals towards which to aim are liberating in terms both of clarity of intent to better frame innovative thinking and also provide flexibility concerning steps progressively navigating institutional practices towards them. This in turn provides a clear basis for communication with wider society about what is being done to progress towards delivering societal needs more deeply rooted in sustainability, and it also serves as a foundation for dialogue with regulators, voluntary organisations and knowledge-providers about establishing relevant milestones and overcoming further obstacles to progress.

Examples below address sustainability issues about which there is already some consensus, if not yet substantial integration of societal energies: circular resource use, 'Net Zero' climate-active emissions, 'Nature Positive' and elimination of persistent toxins (see Boxes 8.1–8.4, respectively). What does it take to bring these missions to life, at least here in illustrative terms? Examples touched upon in this book are drawn upon, but substantial work has to be done to bring societal sectors together to flesh out broad and vivid shared visions from which to work out steps that can lead progressively towards them.

BOX 8.1:   REACHING FOR THE CIRCULAR ECONOMY

Attainment of circular resource use is a challenge beyond the capabilities of any singular societal sector, let alone any individual business, regulatory body, NGO or academic institute. All of us, individually and within the sectors of society within which we primarily operate, have roles to play that interact strongly with others.

• NGO campaigning needs to move from bottom-line attacks on current performance (though this is still warranted for laggards in business and gaps or obstacles in policy) towards working with enterprises seeking to invest in circular resource use as well as influencing the policy and fiscal environment to facilitate its attainment.

- The regulatory sector has the option of working in partnership with committed businesses and business sectors, regulating them in terms of meeting audited voluntary targets around a shared vision of moving from linear to cyclic resource use. This includes revisiting and revising obstructions imposed by legacy regulations, developing inducements including fiscal measures, using the weight of public procurement to promote pro-cyclic products and practices, and developing policies promoting their wider uptake.
- The business sector is the leader in innovation towards circularity in terms of revision of formulations and fabrications facilitating recovery and recycling, as well as investing in take-back and recycling enterprises that are most likely different from the primary manufacturing companies. Developing clear voluntary commitments with audited and reported targets builds trust with regulators, NGOs and wider society, which can engender support for successful innovations and wise investments in cyclic practices.
- The academic and knowledge-providing sector can support progress towards cyclic resource use through the development of tools to determine energy and chemical footprints, optimal locations locally and in globalised markets for recovery and recycling, and consensus-building to bring societal sectors together to address common challenges.

BOX 8.2:   SEEKING THE ATTAINMENT OF 'NET ZERO' CLIMATE-ACTIVE GASEOUS EMISSIONS

Achievement of net zero climate-active emissions is also far from the purview of any single player or sector, requiring a deep societal integration of efforts.

- NGOs and other voluntary groups can develop a championing role to recognise, promote and suggest policies and practices that move incrementally towards net zero. They can also continue to exert pressure on 'climate criminals' as well as obstructive policies and fiscal measures. Harnessing public awareness and promotion of climate-positive choices, and threats arising from failing to do so can also accelerate progress towards carbon neutrality.

• As with the circularity goal, the regulatory sector can choose to work in partnership with committed businesses and business sectors, pivoting the regulatory approach towards the assessment of progress against audited voluntary targets related to the shared vision of the ultimate attainment of carbon and climate neutrality. The 'old school' punitive approach can be reserved for those enterprises that fail to commit to a visionary approach, acting as an inducement for them to join proactive enterprises and the codes of conduct that they develop. Further inducements may include fiscal measures as well as using public procurement to promote products and practices with improving carbon and climate performance. There are also opportunities to refine communications and both regulatory and fiscal policies, relating for example to procurement, to promote consumer and other goods with low lifetime climate impacts. Wider policy reform can accelerate societal progress towards the openly shared goal of attaining climate neutrality.

• For the business sector, the promotion of innovation towards climate neutrality shares many of the facets of the pursuit of circularity. This includes revision of formulations and fabrications to achieve lower whole life cycle energy inputs and embedded carbon. Openly communicated voluntary commitments with reporting and regulatory dialogue about associated audited targets build relationships across society and support for products and services embedded in increasingly climate-aware value chains.

• Knowledge providers have significant roles to play in developing tools to assess life cycle carbon budgets, including raw materials and energy use from all sources (fossil, bio-based, cyclic use) as well as throughout whole product life cycles. There is also a key role to play in dialogic tools to help societal sectors mobilise around common and uniting challenges, and in training future professionals in related disciplines in systemic thinking.

BOX 8.3:   STRIVING COLLECTIVELY
FOR 'NATURE POSITIVE'

Realisation of 'Nature Positive' (regeneration of productive ecosystems) cannot happen unless all sectors pull together to determine how it can be attained within the complexity of the ways in which society operates.

- For the voluntary sector, impacts on nature have been a major campaigning topic since at least the 1970s and for some time before (for example in the founding of the precursor of the Royal Society for the Protection of Birds [RSPB] in 1889 in the UK on the back of concerns for potential extinction of great-crested grebes through their exploitation in millinery[2]). Pro-nature campaigning and awareness-raising are required to engage the wider public, including the policy-making and regulatory sector, and can also help businesses recognise their footprints on biodiversity and seek solutions based both on the sharing of best practices and further innovation.
- The regulatory sector has significant roles to play in developing policies that require recognition, declaration and reduction of footprints on nature, including water resources and soils. The nascent EU Deforestation Directive promises to drive businesses across Europe to audit their supply chains with regard to the physical destruction of nature and the displacement of human rights, promoting innovation of biodiversity-neutral or regenerative practices. Further policy requirements to promote regenerative uses are required, for example stimulation of regenerative agriculture as well as measures such as implementation in the UK in 2024 of mandatory 'biodiversity net gain' in-built developments.
- Business has roles to play in taking greater responsibility for supply chain stewardship to ensure that pressures on biodiversity are reduced to zero and, ideally, progress beyond that to regenerative practices. There is a role for product and service development to ensure neutrality in the life cycle, for example eliminating dependence on large amounts of water or other biologically based inputs in use, as well as greater potential for recovery and recyclability to reduce impacts on nature through disposal.

- The knowledge-providing sector can help by developing robust tools to determine the total footprint on nature of enterprises and the life cycles of products, from supply chains to post-use. It can also research innovative approaches for footprint reduction and, ultimately, progression to regenerative practices.

BOX 8.4:   WORKING TOGETHER IN PURSUIT OF ELIMINATING PERSISTENT TOXINS

Elimination of persistent toxins across whole life cycles may have often formerly been seen as a problem for the producing industry alone. However, all in society use, maintain and dispose of products. Patterns of resource use are affected by understandings, price signals, available infrastructure and marketed choices amongst a range of factors.

- NGOs have roles to play in the 'boxing ring' in exposing and putting pressure on polluters. They also have potential progressive roles in championing benign and beneficial resource uses and products, including those that demonstrate progress towards circularity or that wholly contain potentially problematic materials in tightly controlled manufacturing phases or technical cycles to eliminate exposure and hence risk.
- Regulators need to pivot from a narrow focus on hazard to one that is risk-based, with regulation based on audited targets linked to voluntary commitments that lead towards circularity and elimination of emissions not merely just at manufacturing sites but across whole societal life cycles of products.
- Businesses will be increasingly aware of liabilities associated with emissions of persistent substances, not only at manufacturing sites but produced by products and materials across their whole societal life cycles. This can be a driver towards the circular economy, deselection of the most problematic substances, but also more stringent controls on substances that have societal benefit but need to be better contained across the life cycle. No substance can automatically be assumed to be benign in the context of potential risks across whole societal life cycles, so it is essential to develop wider connections and integration of efforts and investments with other players across society involved in those life cycles.

- Knowledge providers are competent at hazard assessment but need to evolve their approach to assessing risk across societal life cycles, as well as innovation processes allowing safer handling.

All players are deeply interconnected in addressing the many challenges entailed in the ultimate attainment of these and other broad societal goals. It is also evident that all these goals are deeply interconnected. Clearly, greater society-wide intelligence is required to harmonise the energies of all sectors to accelerate their attainment. These shifting roles with respect to reorientation towards a symbiotic approach to address cross-sectoral consensual 'pole star' goals is represented in Figure 8.3.

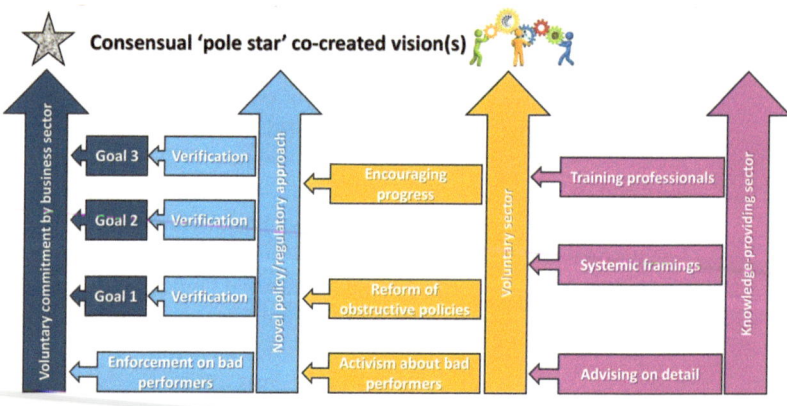

FIGURE 8.3   Shifting societal roles with respect to reorientation towards a symbiotic approach to address cross-sectoral consensual 'pole star' goals.

8.4 NURTURING 'BABY STEPS' TOWARDS A SYMBIOTIC SOCIETY

The examples of synergies between sectors in the development of supply chain stewardship schemes, cyclic recovery of resources after products reach the end of their useful lives, and strategic signals against which innovation can be framed are to be welcomed. The legislative sector is adding rigour to green claims and CSR activities in the corporate sector, leading solution-oriented NGOs are joining multi-sectoral consortia to promote responsible stewardship schemes, and research institutions are shining a light on a pathway in which multiple players have contributory and self-beneficial roles. Objectively, these are tentative 'baby steps' towards a symbiotically functioning society. Though at present hesitant and fragmented, they are nonetheless steps that need to be commended, nurtured, learned from and applied more widely to harness cross-societal energy and intent to accelerate sustainable development.

Germane to this 'baby steps' stage of societal evolution is the saying *"It takes a village to raise a baby"*, a metaphor that becomes graphically evident to new parents. Not literally a physical village, of course, but you rapidly realise when you have a child that you need all the help you can get from family, healthcare professionals, parental networks, friends and neighbours, pre-school and childminders, educators and many more. We are social creatures. Businesses too are social creatures, interactive with a bewildering array of partners – regulators, markets, supply chains, financial institutions, neighbours, staff, unions, competitors and trade associations to combine the interests of ostensibly competing companies and many more – if they are to function efficiently and profitably.

We remain though still shy of serious and substantive change towards sustainability and need the support and engagement of all of this societal 'village' to help the baby grow up and realise its potential. Strategic signalling by governments needs to be backed up by regulatory easements and/or subsidies or other incentives for innovators. More regulatory and NGO acknowledgement and trust is required in response to substantial voluntary, audited progress. Under the VinylPlus programme, delivery against voluntary agreements to electively phase out some problematic substances, invest in and increase post-consumer recycling rates, decarbonise and increase resource and energy efficiency are all progressive steps that deserve recognition. Promotion of a shared international approach will also be required, including quality control of cross-border imports and exports regarding environmental and ethical footprint, not merely as financially based market protectionism but to ensure that less responsible producers are not permitted to undercut those committed, with associated investments, to a pathway of sustainable development. Beyond the establishment of *de minimus* baselines in legal controls within and between nations, we need urgently to progress beyond to create momentum for innovation accelerating sustainable transition while there is still time.

Objectively, we are today a long way from joining hands across societal sectors in a proactive mission to accelerate progress towards sustainable resource use, still mired as we are in inter-business competition and inter-sector antagonism that blunts focus on long-term, consensual goals. This mirrors the conclusions of the 2023 OECD report *Understanding and Applying the Precautionary Principle in the Energy Transition*[3] that proximal concerns about the deployment of renewable energy infrastructure overshadow the greater threats posed by its potential contribution to addressing climate change, potentially existential yet perceived as distant and indistinct. We have some cultural 'growing up' to do to acknowledge the real priorities and the end goals that should be of primary importance in directing our attention and decisions.

Achieving symbiosis across enterprises and societal sectors towards addressing the common goal of sustainable development is itself a vision. Like the end goal of sustainability, it is also still distant. Deeper progress towards sustainable value chains necessarily requires connection across multiple institutions collaborating around common frameworks of understanding and shared goals. This requires links between different institutions in the value chain, through raw material suppliers to material manufacturers, compounders and converters, product users including distributors and retailers, and recyclers or waste-handling interests. However, it must also necessarily entail a significant degree of co-opetition or, in other words, collaboration between potentially competing institutions around common goals to combine their energies and investments to drive progress, whilst allowing for competitive marketing at the product level. Examples here include collaboration to influence the policy environment, pool research investments and promote best practice as well as to petition for statutory enforcement of a 'bottom line' preventing undercutting by less responsible companies including those importing from other global regions with less rigorous environmental and social standards. Trade associations are best placed to represent the common interests of discrete, potentially competing businesses in the value chain. They also provide a vehicle for amalgamating investment, for example in the case of prior examples of companies pooling funds to improve take-back and recycling infrastructure. Without collaboration, this policy leverage and building of infrastructure is unlikely to occur. However, once established, individual companies are then free to compete within an environment more enabling of sustainable progress.

As addressed previously, digital technologies are rapidly evolving and offer substantial potential to accelerate innovations in material types and uses, increase collaboration between societal sectors and develop insights transparently and rapidly about the consequences of material innovation, use and fate. Supporting the 'pyramid model' approach with assistive digital tools could accelerate its uptake across society and its contribution to the sustainable use of materials. These opportunities need to be grasped to reflect the urgent need to be better prepared for a very different future and to address the meeting of societal needs in the face of substantial challenges both in the present and intensifying into a resource-constrained future.

As material use affects outcomes for all societal activities, reforming resource use habits is nothing short of wholesale culture change. This is a transformational journey from deeply entrenched norms and vested interests towards mature engagement with the challenges of an inevitably very different future. We know we can achieve substantive and surprisingly rapid societal transition, as evidenced by global and cross-sectoral collaboration in the face of the Covid-19 pandemic that, not to dismiss its disruption and mortality, is objectively a lesser risk than runaway climate change and biodiversity collapse. We have seen the famous example of the mission to put a man on the moon, and other mission-led 'moonshot' projects breaking barriers of conventional thinking such as innovations in artificial intelligence (AI), robotics, cancer treatments, advanced medical imaging and many more. The journey towards substantive engagement with sustainable development proportionate to the risks we face is long, but we at least have examples of current best practices across all societal sectors heading towards this distant 'moon'.

8.5 HUMILITY AND PATIENCE

A degree of humility is required on the part of governments, regulators and businesses alike, as well as pressure groups and others urging for change, acknowledging that science only gives us imperfect knowledge at any one time. The journey of development, both of knowledge and of material and product innovation will take effort, trial and error from the starting point of today's unsustainable baseline. It will also challenge contemporary norms and vested interests.

It is deeply frustrating to see the perpetuation of unsustainable norms with respect, for example, to continued investment in the exploration and exploitation of fossil fuel reserves, or the abandonment or postponement of climate-related commitments, despite near-global rhetoric about a commitment to decarbonisation. Perpetuation of practices that disrupt aquifers and over-abstract water resources or permit continued emissions of noxious pollutants with grave consequences for aquatic life is equally vexing. We also still too commonly see lax constraints on practices contributing to deforestation despite proclamations to the contrary, as well as the dispossession of native people through appropriation of land, landscapes and mineral and biological resources. As a life-long proponent and campaigner agitating for environmental responsibility and sustainable progress, these continued offences against nature and people in pursuit of quick profit verge on and can veer into actual psychological pain and grief.

Concerned people will, as I do, continue to urge all parties to accelerate change. However, it is also important to acknowledge that immediate attainment of sustainability is practically impossible in the short term, as we saw with the example of how unrealistic and indeed detrimental it is to expect or demand instant attainment of the end goal of the EU's 'Clean chemistry' and

'Circular economy' strategies. Humility and patience are required to recognise and congratulate incremental, practically attainable and profitable steps on the strategic journey towards the ultimate attainment of distant goals. Clearly articulated strategic goals are vital, supported by committed and audited progress towards them with the knowledge that these represent way-markers on a journey towards sustainable destinations. We owe this degree of commitment and more to future generations. Let us also not forget that we owe it to ourselves too in terms of the way an unsustainable economy ultimately damages our life expectations including our pension plans!

Whilst advocating patience, it is equally important not to rest on yesterday's successes. As discussed previously, yesterday's shocks become embedded into the mainstream of business, societal and regulatory trends, and eventually constitute new norms. Today, we accept restrictions on many substances with known toxic properties, slavery and nuisance to neighbours of production facilities, smoking in public places, speed limits on roads and the banning of lead in road fuels, all of which represent novel innovations from decades past. We need to accept that yesterday's voluntary commitments by business will, if relevant, become progressively embedded in regulatory requirements and public and market acceptability such that they are no longer voluntary. Some NGO campaigning successes about chemicals and resource use practices too become mainstream issues, and so it is necessary for campaign groups to move on to agitate for more progression towards sustainable end goals rather than repeat anachronisms that only serve to stimulate defensive behaviours in business. Society gradually absorbs and establishes new norms in the journey upwards and onwards towards its pole stars and needs to renew its vows in terms of new voluntary and proportionate commitments.

As we look into the future, it is increasingly and objectively clear that the world is facing a raft of unavoidable and potentially existential threats. It is also objectively clear that tomorrow cannot be a simple extension of today but will certainly be disrupted by these interactive pressures. On that new, unstable horizon, the ways in which we conceive, innovate and use materials will constitute a significant element in addressing continuing societal needs. This is a mission we all share.

8.6 SYMBIOSIS BEYOND MATERIAL USE

The 'pyramid model' presented here articulates the spectrum of fragmented and potentially divisive potential characteristics of the four principal sectors of human societies – private, public, voluntary and academic – as well as their potential for co-creative and mutually supportive action. We have discussed this in terms of driving towards the sustainable use of materials, but it is also relevant to all other sectors of human activity. Four illustrative examples of the wider applicability of the 'pyramid model' are given in Box 8.5.

BOX 8.5:   APPLICATION OF THE 'PYRAMID MODEL' TO FOUR ILLUSTRATE DIVISIONS OF HUMAN ENDEAVOUR

In healthcare: global mobilisation in response to the Covid-19 pandemic taught us how society can genuinely pull together when grave threats focus collective societal attention on evident needs to address common goals. Governments collaborated to rapidly develop quarantine processes as well as prioritising and pooling investment in vaccine development and roll-out. Businesses were very much part of this process through innovation, manufacture and supply of personal protective equipment and development of vaccines, in many cases in close partnership with knowledge-providers from the academic sector. In many regions around the world, NGOs mobilised to support communities, particularly the most vulnerable.

In the management of river catchments: the voluntary sector has been active in focusing attention on integrated outcomes to protect or improve river health, for example in the case of the network of Rivers Trusts across the British Isles.[4] Amongst the activities of this network of catchment-based Trusts is engagement with the academic sector to commission research and to inform best practice. There is also a handshake with regulators, attempting to better direct fragmented regulatory duties and budgets to achieve integrated outcomes for the distinct characteristics and challenges of local river systems. Some businesses sponsor these improvement ventures, while others, particularly water and sewerage companies, are perceived as requiring greater commitment and investment to meet their environmental obligations. Water service companies though are also major beneficiaries of improved river quality as water abstracted for public supply then requires a lower level of treatment. Farm businesses too are key actors both in diffuse pollution and other pressures on river systems but also in the implementation of protective processes and infrastructure.

In urban and community planning: the government sector has obligations towards the sustainability of urban and community planning with an increasing interest in liveability. NGOs and other civil society organisations are often active in engaging people in consultation processes, raising objections where infringements are perceived, and in making suggestions for improvement. The academic sector comprises significant expertise in urban design relating to the physical landscape but also mechanisms to facilitate institutional dialogue. It is businesses that get engaged in construction, whether influenced by a collective vision or simply maximising profit.

In defence: whilst it may not be initially intuitive to consider contributions to sustainability, visions converge in the aspiration to achieve peaceful coexistence including both peace-making and peacekeeping. This goal is aspirational for governments, which have roles in resource allocation and upholding of rights. Many NGOs seek to build or maintain bridges within divided communities or, in the worst case, support social and physical reconstruction after conflict. Academia provides knowledge about the causes of conflict and can inform social dialogic practices to heal rifts or maintain harmony. Businesses provide the hardware and software supporting these efforts and are also key actors in reconstruction.

We live at a pivotal time in human history wherein the magnitude of threats arising from what we might assume to be acceptable norms is threatening to undermine the supportive capacities of the planetary ecosystems and the rights of all who share them, now and onwards into the future. We can continue on this trajectory and seek to maximise profit and individual wellbeing on a short-term, selfish and competitive basis, or we can alternatively look ahead and agree that collective commitment and action are required to deliver a world that continues safely, securely and indefinitely: sustainable development.

NOTES

1 Everard, M. (2024). *Seeking Sustainable Development on a Level Playing Field: A PVC Case Study*. CRC Press, Boca Raton, FL.

2 RSPB. (2024). *Our History.* Royal Society for the Protection of Birds (RSPB). [Online.] https://www.rspb.org.uk/about-us/our-history?gad_source=5&gclid= EAIaIQobChMI1_-op6XFigMVFZhQBh2YTRKUEAAYAiAAEgKDrvD_ BwE, accessed 26 December 2024.
3 OECD. (2023). *Understanding and Applying the Precautionary Principle in the Energy Transition.* OECD Publishing, Paris. https://doi.org/10.1787/5b14362c-en.
4 The Rivers Trust. (2025). *The Rivers Trust.* [Online.] https://theriverstrust.org/, accessed 23 January 2025.

A View of the Journey from the Future

9

Future generations will judge us for the actions and decisions we make now, including our wilful or casual inactions. The same is true of today, as I write from the middle of the twenty-first century, as we live with the legacy of decisions and committed actions or oversights of preceding generations.

Looking back, it has been quite a journey to where we stand today. Looking forward, the journey to where we know we need to reach is at least as long; most probably a lot longer as we navigate towards a rather differently rooted future. Over the past few decades, we have travelled a rocky path, jolted from yesterday's lazy habits by resource price hikes, negative press and campaigns that generated public and market hostility, the combined weight of which has made our board sit up and listen. Legislation too seemed to ramp up in response, aided and abetted by vocal NGOs that, looking back now, expanded our narrow awareness of our slothful procurement and manufacturing practices and our disregard for where our products ended up. Whilst some countries rowed back on their environmental and social responsibilities and prior commitments in blind pursuit of naked competitive profit, and are now clearly viewed as pariah states with which almost no-one wants to trade, we held to our commitments and have done well as a result. It is crazy now to think that the liabilities and reputational impacts stemming from irresponsible use of what we made and sold to others would not come back to blight us.

From the first quarter of this century, awareness dawned amongst the wider global community that we are all in this together. What we make, how we invest and the things we profit from are, in reality, just part of a far bigger dynamic. We have found that doing what we do with integrity and concern for consequences – making things that serve clear needs without unacceptable risk – offers secure profitability; this is a world apart from our former habit of trying to sell more and more stuff into a volatile market, with increasing 'naming and shaming' with respect to outcomes we had not foreseen or had perhaps turned a blind eye.

174

DOI: 10.1201/9781003637875-9

9.1 WE ARE ALL IN IT TOGETHER

The biggest shift has been in our relationships with others.

With our competitors, we work largely via trade associations to pool effort and investment over those things we cannot do alone – dialogue with regulators, public perception of our sector, joint research into groundbreaking but expensive new technologies and proofs of principle, and more besides – to create space within which we then compete with more sustainable yet differentiated products.

We also collaborate more closely with our customers, their customers and the infrastructure that recovers resources when products have passed their useful lives. Why on earth did we ever think that merely waving goodbye to our products as they left the factory gates was good business? The reality, of course, is that every player in these value chains is in this together, either serving the needs of people as efficiently and safely as possible or else creating chemical, physical and social problems as well as liabilities down the line. *"Doing well by doing good"* may have been an old rallying cry, but we have found that it is one of those truisms that just happens to be true… No surprise there, I guess!

We have close relationships also with our suppliers. Long gone are the days when we continued to buy from companies supplying our raw materials, packaging, energy, transport and other inputs who were not prepared to disclose their footprints. Application of blockchain to transparently cryptographically code elements of total value chains and to inform smart contracts has really helped. We formerly got fed up with NGO and media activism revealing that we were unknowingly complicit in unethical and environmentally destructive trade, the criticism stung but also shamed us and dented the trust of our customers. So now, open disclosure is a new normal, one upon which we insist though also now something that is increasingly demanded by regulations and is clearly demonstrable digitally. Regulations now rightly also extend to what is allowed to be imported from more opaque global regions, not as a trade barrier but as a quality control. As I have said before, *"Doing well by doing good"*.

Shifting relationships with regulators has been particularly enlightening. Back in the old days, our regulators were principally concerned just with 'bottom line' performance, imagining that we could somehow incrementally ramp up compliance with tighter environmental and social standards in a way that radically changed established societal norms and vested interests. However, realisation dawned that sustainable development meant something rather different than merely 'being less crap tomorrow'. Rather, it required measures actively promoting or enabling substantial innovation in the face of daunting challenges in a world of limited resources but escalating needs. This shifting awareness drove government and the regulatory community to recognise the importance of pairing bottom-up regulation with something that was genuinely aspirational and vision-led. Now we have strategies for product sectors and industry types that are not just imposed from above but are co-created with regulatory communities. Our part of the deal, beyond co-creating articulate visions, is then to commit, largely via trade associations representing sectoral interests, to voluntary strategies with associated audited targets working towards those longer-term shared visions. It is this audited progress that the regulators now primarily look to as a means for ensuring that we, and society at large, are taking tangible steps in the right direction. The relationship has substantially shifted towards one of partnership based on a common vision about how we can move together and with increasing speed to create products and use materials in increasingly safe and efficient ways that contribute better to clearly articulated human and environmental needs.

Our relationship with voluntary organisations and media has also changed significantly. Campaigning NGOs still nibble at our toes, and rightly so in some cases because some businesses undoubtedly require greater scrutiny regarding negative behaviours, opaqueness or failure to grasp or engage with progressive moves, and these laggards pose reputational risks to us all. However, many in the NGO movement have been shifting since the 1990s towards more of a solutions-based orientation. That trend has continued as they collaborate with us on the development of common visions regarding how the use of materials, products and technologies can better support meeting needs. Some of these leading NGOs are also important knowledge providers and stimulants as well as scrutineers of progress; a welcome relief from the constant criticism we often formerly encountered despite making substantial investments to move in the right direction.

Researchers, both in academia and in consultancies as well as independent experts, are part of this more collegiate approach in the building and then practical delivery of shared visions. They give us insight but also keep us honest about what is wise and what is foolish when it comes to making progress. Their training of future professionals is also more rounded in terms of the contributions that each discipline can make towards sustainable development, making hiring of appropriately skilled and visionary staff less of a lottery than back in the bad old days.

We also have a changed relationships with our customers. This includes both those downstream in the value chain in manufacturing, product development and product use and, by extension, also with the consumers of finished products and onwards to those recovering and recycling products reaching end-of-life. Green consumerism is alive and well, is more transparently enabled by computer systems and constitutes an important 'pull factor' in markets. But we also recognise a responsibility that we should not offer consumers flawed products. So, for example, across the co-created visions, we actively deselect the application of some materials in certain markets where other materials can serve needs more sustainably while being up-front in emphasising the benefits of what we produce in terms of optimally meeting needs safely and efficiently in the product sectors into which we continue to supply. We also want to ensure that our customers are not presented with obstacles to benign disposal or recovery at the end of life of our products. In that regard, objective demonstration of greater sustainability has become a significant market differentiator.

9.2 SHARED VISION

I've spoken on and off in this narrative about the visions that we co-develop. A key facet of this is that these visions are not related solely to today's problems. Rather, they are grounded on clearly articulated, science-based frameworks covering all dimensions of sustainability and looking at whole product life cycles. We co-create these visions across companies within our various business sectors, with regulators and also with input from researchers and voluntary organisations. The key skill behind the creation of visions is that we go way beyond simply defining problems, which tends to be debilitating, looking instead at where we need to get to in the long run. It is not a question of being less bad, but of articulating in as graphic terms as possible how to achieve the longer-term meeting of sustainability goals and recognise the roles of all players in their attainment.

The co-creative element is critically important. Everyone brings their interests to the table, but we try to look beyond factional positions. In the early days of the twenty-first century, there were clamours to *"Get rid of plastic"*! Yet, working together, it was quite clear that plastic itself was not the problem, but that accumulation of waste was the core issue. So, creating a backcasting position, and acknowledging leadership on this matter by CSIRO in Australia and other global NGOs, we worked together to set and agree on a goal that eliminating plastic waste was the place we needed to get to. And how could we achieve that? Well, it clearly wasn't just the producers that were responsible, even though they had formerly shouldered most of the blame. After all, plastics in wide societal use were there for many beneficial reasons, including, for example, keeping food fresh, serving as durable infrastructure in the built environment, medical and healthcare products, and many other beneficial contributions. Many sectors of society were benefiting from the wise use of plastics but, equally, some were also contributing to the unwise use of these materials. So, the vision included the progressive deselection of plastic from short-life and particularly single-use applications where containment into recovery systems was unlikely, representing a wasteful and waste-generating use of valuable materials, but the retention of plastic applications where backed up by achievable take-back and cyclic reuse. We then focused on where these durable materials could add the most value, particularly in the built environment and healthcare, working across societal sectors to collaborate on how different players could contribute to recovery and recycling infrastructure. We also looked at which types of plastic were most suited to specific applications. Other examples included wiser use of metals as well as glass, ceramics and the wide diversity of other materials in common use in the world. It was vital that we examined this from the perspective of a 'level playing field' of common principles, rather than making prejudgments and defending what we thought were 'good' or 'bad' materials considered in isolation from their practical use across the life cycles of 'real-world' products.

We also recognised that we live in a non-linear world, meaning that back-casting had to be done cautiously. Acknowledging that there was no simple linear pathway from where we stood, and still stand, regarding unsustainability and onwards to the ideal of a fully sustainable future, there will clearly be many bends and kinks in the path necessitating a constant review of the vision. For this, we maintain working groups populated by regulators, trade associations representing common business interests, as well as NGOs and other players. These working groups help us stay current in the face of emerging trends, remain on the right track, keep us honest and make sure that the voluntary commitments that we make are regularly updated, audited appropriately and dovetailed with the very different regulatory frameworks under which we operate today when compared with the old punitive model.

The world of shared visions is so much more liberating and empowering, and so much less wasteful, than the ways in which we operated at the turn of the century. There is a sense of teamwork across society as we stare in a united way into the face of daunting challenges presented to us by an inevitably different future, and as we work out together how best to navigate towards optimal and mutually beneficial outcomes.

9.3 UTOPIA?

People said that serious commitment to this journey was a pipe dream: we were chasing an unattainable utopia. Is that true?

One thing that became ever clearer was that failing to work together towards a commonly shared vision was to guarantee division and dystopia.

And, in that darker world, our business would likely not survive if it simply perpetuated what we now clearly see were the bad habits founded on turning a blind eye to the consequences not only of our actions but also of sales of poorly scrutinised products.

Does where we are today feel like a utopia? No way! We still have spats and skirmishes: with competitors, suppliers and customers, about fair shares of investment in joint research and in the recovery and recycling infrastructure, as well as with how regulators scrutinise audited reports on progress with targets set at company and business levels that mark stepping stones towards voluntary commitments. We have sporadic friction also with NGOs and media demanding we go faster and further than is affordable, or when they obsess over fine details but forget the longer-term mission that we all signed up for together.

The thing that keeps us all on track is that, at a higher level, we share the vision of where we need to get to and the commitment to progress towards it over time. This is great for staff morale, retention and motivation, as our people feel they are part of a greater purpose than just putting more money into the pockets of our shareholders. Effectively, we operate as an extended family of common interests and, let's be honest, I have yet to encounter a truly utopian family that does not have tussles over minutiae despite the strength of their common bond!

9.4 THE JOURNEY AHEAD

Reflecting backwards, any serious and committed observer would doubtless have concurred that, for all the high rhetoric in those first decades of the century, the pace of societal engagement and practical progress with sustainable development was lamentably shy of the deepening threats posed by stark evidence of trends in climate instability, biodiversity decline, water stress, resource security and inequalities.

It is this convergence of threats, and how they began to impact us increasingly more deeply, that helped us recognise the urgent need to pull together around focal and consensual visions and proportionate action plans to reform our lax habits, with the shared promise of improving our collective security and life opportunities.

The journey ahead is long and challenging. But, comparing the present situation with where we were in those awakening days of the 1970s, it is amazing to see how far we have come. Back in the dawning days of the modern 'environment movement', and its segueing into awareness about the need for a broader sustainable pathway of development, we hardly considered our negative impacts on people and the planet. When a light was shone upon them, we most commonly went into denial or post-rational justification. It is amazing that any of us survived that darkly destructive era of our past!

We come to work now, and make and sell things, with a purpose greater than just contributing to the net worth of shareholders. It feels good to see how what we do now, and what we aspire to do increasingly better in collaboration with all these other organisations, adds value to people, now and tomorrow. We know that the world is far from perfect – after all, today's norms are still tainted by the legacy of where we have been – but we now at least share a cross-societal acknowledgement and commitment to a common vision of serving society's needs safely and efficiently.

We need to step beyond this too, recognising that, however visionary our thoughts may be, we inhabit a material world. Dreams and aspirations that are not incarnated into some physical or social structure remain unfulfilled. We build the future in one way or another with materials, and so the diverse ways in which we develop and use materials are central to all strands of sustainable development and the meeting of needs both now and tomorrow.

Accelerating Towards Sustainable Use of Materials

10

The progressive clothing of visions and novel ideas into material form has driven the rise of technologies, revolutions and cultures and formed the metaphorical, or sometimes literal, bedrock upon which entire civilisations have been built. Cascading innovations throughout history and the material ways in which they have given physical form to more efficiently meet our diversity of needs and whims have improved life prospects. Many of us in the already-developed world today enjoy unprecedented levels of nourishment, medical care, mobility, entertainment, comfort, wealth and ease. We, the lucky upper percentile, stand as beneficiaries of humanity's cumulative intellectual inheritance in the manipulation of the natural world.

We also potentially stand on the precipice of its decline, as the ways in which we have exploited materials also cast a darker shadow. Material choices and use patterns have often also been a key contributor to subsequent decline and vulnerabilities through resource exhaustion by over-exploitation and our ever-greater technological reach, competition for dwindling supplies potentially descending into conflict as well as pollution and reductions in the vitality of supporting ecosystems. Limitation of vision about potential consequences has led us to build a technosphere that provides a cornucopia for the privileged. However, it is also one that not only marginalises the least powerful in society but also ultimately consumes itself. The pace and extent of our use of soils, biodiversity, minerals, water, the atmosphere, the chemicals of which these media are composed and the substances that we synthesise from them is now raising barriers to further progress. Worse still, some

DOI: 10.1201/9781003637875-10

of our material use practices are generating existential threats and imposing an unavoidable blight on future generations.

At this point in human history, burgeoning human numbers and demands are facing a conflict with dwindling resources. For this reason, there is a pressing need for concerted societal efforts towards sustainable development. We also now have the scientific knowledge to understand interdependencies both with supporting ecosystems and those with whom we share them. The word 'system' is the root of the words 'ecosystem' and 'socio-ecological system', emphasising how humanity is integrated locally, within river catchments, across land masses and right up to global biospheric scale. Complex feedback loops occur from all our interactions with supportive ecosystems and those who share them as common resources, for better or for worse depending on whether we work divisively or pull together around common goals. At this precarious crossroads in human history, there may be no more important collective goal than concerted progress towards sustainability.

We can choose to act antagonistically across society, dissipating our creativity and energies, or else take a step back to look at aspirations upon which we can all agree and work towards symbiotically. The 'pyramid model' presented in Chapter 8 articulates the spectrum of fragmented and potentially divisive potential characteristics of the four principal sectors of human societies – private, public, voluntary and knowledge-providing – as well as their potential for co-creative and mutually supportive action.

Whilst developed in the context of this book for the challenge of working towards the sustainable use of materials, it is also relevant to and can support challenges in all other spheres of human activity as indicated in the wider examples explained in Section 8.6.

A consensual vision is a spur for collaboration across society to work towards necessary innovation to better meet human needs in a fast-changing world. This transformative approach begins by asking what need is being addressed as a first principle, secondly how this can be served in the safest and most efficient manner and thirdly about how humanity can work symbiotically across different interest areas to attain mutually beneficial outcomes.

Attainment of a level playing field of material assessment, regulation and innovation is a key component of this. It will be achieved by taking a 'material blind' approach, seeking the best, most efficient and safest means and material choices or innovation to address needs on a context-specific basis. Today and tomorrow, the supportive capacities of the world are simply not sufficient to fulfil the growing and diversifying needs of humanity unless we radically redress current habits, assumptions and trends that are already placing intolerable pressures on fast-dwindling ecosystems and resources. Future profitability will necessarily be increasingly framed by enduring value to humanity, rather than rapid capitalisation that places immediate profitability over and above the satisfaction of needs and longer-term, linked consequences for ecosystems and people. This is a uniting mission within which we are all co-beneficiaries or victims: we are unavoidably all in it together. For this big mission, we need a big and connected society working symbiotically, co-creating visions that unite efforts for well-understood benefits in a future that is more wholesome and secure, and with expanding rather than constrained opportunity for all. This is true locally, regionally, nationally and also internationally. Just as sustainable development cannot be realised by any institution alone, the same principle applies equally to all nations within a world intimately interconnected not just by globalised markets but within a single shared biosphere. We are all ultimately bound by the same natural laws and workings of the living planetary system, within which we and all our activities are interactive elements. We all therefore ideally need to share the same understanding and vision of attaining a sustainable world, telling ourselves vivid but locally nuanced stories of meaning to inspire and accelerate the attainment of greater security and fulfilment.

Accelerating societal progress towards the sustainable use of materials is no mere 'nice to have'. Rather, it is the material foundation of not only shared challenges but also how we go about creating a future of greater security and opportunity for all.

Index

Note: **Bold** page numbers refer to tables and boxs, *italic* page numbers refer to figures.